METHODS IN MOLECULAR BIOLOGY

Series Editor
John M. Walker
School of Life and Medical Sciences
University of Hertfordshire
Hatfield, Hertfordshire, UK

For further volumes:
http://www.springer.com/series/7651

For over 35 years, biological scientists have come to rely on the research protocols and methodologies in the critically acclaimed *Methods in Molecular Biology* series. The series was the first to introduce the step-by-step protocols approach that has become the standard in all biomedical protocol publishing. Each protocol is provided in readily-reproducible step-by-step fashion, opening with an introductory overview, a list of the materials and reagents needed to complete the experiment, and followed by a detailed procedure that is supported with a helpful notes section offering tips and tricks of the trade as well as troubleshooting advice. These hallmark features were introduced by series editor Dr. John Walker and constitute the key ingredient in each and every volume of the *Methods in Molecular Biology* series. Tested and trusted, comprehensive and reliable, all protocols from the series are indexed in PubMed.

eQTL Analysis

Methods and Protocols

Edited by

Xinghua Mindy Shi

Temple University, Philadelphia, PA, USA

Humana Press

Editor
Xinghua Mindy Shi
Temple University
Philadelphia, PA, USA

ISSN 1064-3745 ISSN 1940-6029 (electronic)
Methods in Molecular Biology
ISBN 978-1-0716-0025-2 ISBN 978-1-0716-0026-9 (eBook)
https://doi.org/10.1007/978-1-0716-0026-9

This Humana imprint is published by the registered company Springer Science+Business Media, LLC, part of Springer Nature.
The registered company address is: 233 Spring Street, New York, NY 10013, U.S.A.

Preface

Overview

Data tsunami in biology and life sciences has significantly changed the way research is conducted. Scientists are faced with mining large datasets from different sources to answer a particular question. In this scenario, data analysis and information integration become an essential tool in biomedical research. The availability of genomic data at multiple layers makes it possible to co-analyze them in a common framework for new insights. In recent years, quantitative trait loci (QTL) analysis and expression quantitative trait loci (eQTL) analysis have gained momentum due to its power of integrating genetic variation with molecular phenotypes like gene expression. However, due to the complexity of the data and analysis framework, the science community demands a comprehensive protocol suitable for the general audience of biologists and bioinformaticians. In this book, we plan to write such a handbook on state-of-the-art eQTL analysis, where interdisciplinary researchers will provide both theoretical and practical guidance to eQTL analysis and interpretation. We believe such a book will come at the right time as more genomic and transcriptomic sequencing projects are underway to produce datasets for eQTL and QTL analysis on many organisms of interest. Hence, this book is targeted for biologists and bioinformaticians who generate data and provide eQTL and QTL analysis.

Organization

This book consists of 17 chapters which are organized into 2 parts: Part I, Foundations, and Part II, Applications. Each chapter is on a different topic contributed from domain scientists and researchers.

Part I, Foundations, includes 11 chapters that describe foundations, methods, and tools for eQTL and QTL analysis. Chapter 1 introduces commonly used methods for eQTL analysis including approaches to correcting for data biases. Chapter 2 describes R packages for eQTL and QTL analysis, with a focus on demonstrating the usage of a particular tool called genomic tools. Chapter 3 presents a different strategy for eQTL analysis by considering transcription factor binding affinity. Chapter 4 describes state-of-the-art approaches for detecting splicing quantitative trait loci that can be applied to other types of QTL analysis including eQTL analysis. Chapter 5 describes a genome-wide composite interval mapping for eQTL analysis for backcross or doubled haploid populations.

Chapter 6 demonstrates a machine learning-based method for leveraging eQTL information to prioritize candidate genetic variation in trait association regions. Chapter 7 reviews emerging statistical and machine learning methods for eQTL analysis including sparse regression models based on penalization. Chapter 8 proposes sparse regression models identifying group-wise and individual-level eQTL associations in eQTL analysis. Chapter 9 introduces a novel Bayesian approach for eQTL analysis using probabilistic programming. Chapter 10 proposes a new statistical test for eQTL analysis by incorporating both mean and high-order effects of genetic variation on gene expression. Chapter 11 introduces existing methods for eQTL analysis by integrating multi-omics data and describes a new method for eQTL analysis by including the epistasis among multiple genetic variants with regard to their nonlinear effect on gene expression. Chapter 12 proposes a sparse partial least square approach and a package called Matrix Integration Analysis for identifying modular patterns in eQTL analysis and omics data analysis.

Part II, Applications, consists five chapters that further demonstrate the usage of eQTL analysis in various scenarios. Chapter 13 presents a workflow for comprehensive eQTL analysis in cancer genomics to understand how genetic variation contributes to tumorigenesis and development. Chapter 14 reviews different types of QTL analysis using molecular traits in addition to eQTL analysis. Chapter 15 describes data generation, preparation, and the performance of QTL analysis that can be applied to many species toward discovery of candidate genes or loci. Chapter 16 introduces a procedure for multi-tissue eQTL analysis and demonstrates how to prepare data and conduce the analysis. Chapter 17 further demonstrates eQTL analysis in different tissues and describes tissue-specific eQTL in zebrafish.

Acknowledgments

We are deeply grateful for the contribution from 33 authors of the 17 chapters of this book who had tirelessly worked on writing and editing of their contributed chapters. We would like to mention that the responsibility of the contents of individual chapters of this book lies with the concerned authors. We are extremely thankful to the Springer publishing teams, particularly, John Walker, who worked with us and guided us all through every step of the publication of this book. We appreciate Junjie Chen's help in organizing the chapters. We thank our families, friends, and colleagues for their continuous support.

Philadelphia, PA, USA ***Xinghua Mindy Shi***

Contents

Part II Applications

Contributors

JINYU CHEN • *NCMIS, CEMS, RCSDS, Academy of Mathematics and Systems Science, Chinese Academy of Sciences, Beijing, China*

JUNJIE CHEN • *Department of Bioinformatics and Genomics, College of Computing and Informatics, University of North Carolina at Charlotte, Charlotte, NC, USA*

WEI CHENG • *NEC Laboratories America, Inc., New York, NY, USA*

KIMBERLY P. DOBRINSKI • *The University of Tampa, Tampa, FL, USA*

DANIEL FISCHER • *Applied Statistical Methods, Natural Resources Institute Finland (Luke), Jokioinen, Finland*

JEAN GAO • *Department of Computer Science and Engineering, University of Texas at Arlington, Arlington, TX, USA*

JING GONG • *Hubei Key Laboratory of Agricultural Bioinformatics, College of Informatics, Huazhong Agricultural University, Wuhan, P. R. China*

ELENA GRASSI • *Department of Molecular Biotechnology and Health Sciences, Molecular Biotechnology Center, University of Turin, Turin, Italy*

LENG HAN • *Department of Biochemistry and Molecular Biology, The University of Texas Health Science Center at Houston McGovern Medical School, Houston, TX, USA*

MINGON KANG • *Department of Computer Science, University of Nevada, Las Vegas, Las Vegas, NV, USA*

GEN LI • *Department of Biostatistics, Mailman School of Public Health, Columbia University, New York, NY, USA*

YANG I. LI • *Section of Genetic Medicine, Department of Medicine, University of Chicago, Chicago, IL, USA; Department of Human Genetics, University of Chicago, Chicago, IL, USA*

YAOMING LIU • *State Key Laboratory of Ophthalmology, Zhongshan Ophthalmic Center, Sun Yat-sen University, Guangzhou, P. R. China; Department of Biochemistry and Molecular Biology, The University of Texas Health Science Center at Houston McGovern Medical School, Houston, TX, USA*

ZHENG LIU • *School of Electrical Engineering and Computer Science, Oregon State University, Corvallis, OR, USA*

ELISA MARIELLA • *Department of Molecular Biotechnology and Health Sciences, Molecular Biotechnology Center, University of Turin, Turin, Italy*

CONOR NODZAK • *Department of Bioinformatics and Genomics, College of Computing and Informatics, University of North Carolina at Charlotte, Charlotte, NC, USA*

WEIWEI OUYANG • *Parkland Center for Clinical Innovation, Dallas, TX, USA*

KARA E. POWDER • *Department of Biological Sciences, Clemson University, Clemson, SC, USA*

PAOLO PROVERO • *Department of Molecular Biotechnology and Health Sciences, Molecular Biotechnology Center, University of Turin, Turin, Italy; Center for Translational Genomics and Bioinformatics, San Raffaele Scientific Institute IRCCS, Milan, Italy*

HUAIZHEN QIN • *Department of Epidemiology, College of Public Health and Health Professions and College of Medicine, University of Florida, Gainesville, FL, USA*

STEPHEN A. RAMSEY • *Department of Biomedical Sciences, Oregon State University, Corvallis, OR, USA; School of Electrical Engineering and Computer Science, Oregon State University, Corvallis, OR, USA*

ANKEETA SHAH • *Committee on Genetics, Genomics, and Systems Biology, University of Chicago, Chicago, IL, USA*

XINGHUA SHI • *Department of Bioinformatics and Genomics, College of Computing and Informatics, University of North Carolina at Charlotte, Charlotte, NC, USA*

DIMITRIOS V. VAVOULIS • *Department of Oncology, University of Oxford, Oxford, UK; The Wellcome Centre for Human Genetics, University of Oxford, Oxford, UK; NHS Translational Molecular Diagnostics Centre, Oxford University Hospitals, Oxford, UK; NIHR Oxford Biomedical Research Centre, Oxford, UK*

WEI WANG • *Department of Computer Science, University of California, Los Angeles, Los Angeles, CA, USA*

BENJAMIN WEEDER • *School of Biological and Population Health Sciences, Oregon State University, Corvallis, OR, USA*

JIA WEN • *Department of Bioinformatics and Genomics, College of Computing and Informatics, University of North Carolina at Charlotte, Charlotte, NC, USA*

YAO YAO • *School of Electrical Engineering and Computer Science, Oregon State University, Corvallis, OR, USA*

YOUQIONG YE • *Department of Biochemistry and Molecular Biology, The University of Texas Health Science Center at Houston McGovern Medical School, Houston, TX, USA*

SHIHUA ZHANG • *NCMIS, CEMS, RCSDS, Academy of Mathematics and Systems Science, Chinese Academy of Sciences, Beijing, China*

XIANG ZHANG • *College Information Sciences and Technology, The Pennsylvania State University, University Park, PA, USA*

YUAN-MING ZHANG • *Crop Information Center, College of Plant Science and Technology, Wuhan, P. R. China*

JINYING ZHAO • *Department of Epidemiology, College of Public Health and Health Professions and College of Medicine, University of Florida, Gainesville, FL, USA*

Part I

Foundations

Chapter 1

Introductory Methods for eQTL Analyses

Conor Nodzak

Abstract

Expression quantitative trait locus (eQTL) analysis has proven to be a powerful method to describe how variation in phenotypes may be attributed to a given genotype. While the field of bioinformatics and genomics has experienced exponential growth with modern technological advances, an unintended consequence arises as a lack of a gold standard for many applications and methods, which may be compounded with ever-improving computational capabilities. Researchers working on eQTL analysis have at their disposal a multitude of bioinformatics software, each with different assumptions and algorithms, which may produce confusion as to their respective applicability. In this chapter, we will introduce eQTLs, survey commonly used software to conduct a mapping study, as well as provide data correction methods to avoid the pitfalls of such analyses.

Key words Bioinformatics software, eQTL analysis, Linkage and association mapping

1 Introduction

The twenty-first century geneticist has continued to advance our understanding of life's double-helical building block based on the explosive progress in the knowledge base since the structure of DNA was first solved. Although the perspective of molecular biology is by definition dealing with the very small, modern researchers find themselves able to turn to the question of genetic regulation at the genome-wide scale. Experiments designed to study the impact of parental phenotypic variation on heritable traits have their roots in the work of Mendel, and with the technological arsenal of today it is has become feasible to investigate the totality of genetic sequence variation within pedigrees and across disparate populations.

In parallel, the entire transcriptome of an organism may also be probed or directly sequenced and mapped to respective genomic loci in order to quantify the activity of all genes. By using the expression of mRNA transcripts for genes as a phenotype of interest, one may quantify on a continuous scale the effect of regulatory

Xinghua Mindy Shi (ed.), *eQTL Analysis: Methods and Protocols*, Methods in Molecular Biology, vol. 2082,
https://doi.org/10.1007/978-1-0716-0026-9_1, © Springer Science+Business Media, LLC, part of Springer Nature 2020

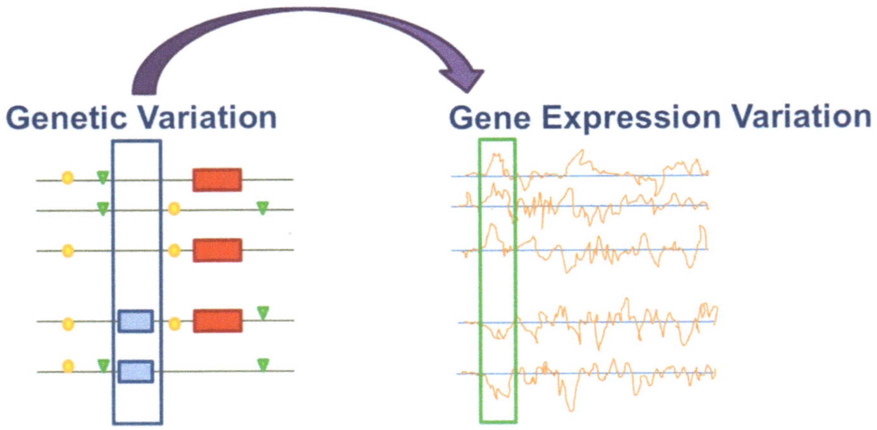

Fig. 1 An illustration on the general idea behind an eQTL analysis. In order to study the effects of genetic variation, pairwise associations are made by correlating the genetic variants in a given population, given by the blue box on the left, to measured expression levels in the green box on the right

variation by calculating the correlations between genetic differences and respective transcript abundance or probe intensity values. This idea is illustrated by Fig. 1, where the genotype variants among samples are linked to their expression levels. Classically, the process seeks to identify these expression quantitative trait loci (eQTLs) as certain variable regions highly correlated with observed differences in gene expression. Typically, eQTLs may be annotated as single nucleotide polymorphisms (SNPs), large structural variants, or copy number variants.

The relative location of the eQTL to the gene may be important with regard to the effect observed. The arbitrary distance-based distinction of eQTLs depends upon the author's discretion; however, a window of ±1 Megabase of genomic sequence may be a common practice in the literature [1]. As a general definition, *cis*-eQTLs are those that lie within a predefined window upstream or downstream of the target gene, whereas *trans*-eQTLs would denote any loci outside the same window and may even be located on different chromosomes. In some cases, one may also be interested in the allele-specific effects of *cis*-eQTLs through a subsequent analysis of the differences among RNA-seq read counts that contain each allele at heterozygous sites.

Genome-wide mapping for eQTLs in this manner lends itself to multiple applications, from a controlled pedigree with limited recombination events to a large and heterozygous population-level analysis. These two complementary experimental designs are referred to as linkage-based eQTL mapping and association eQTL mapping, respectively. The different approaches to experimental design have several strengths for mapping eQTLs; however, limitations exist for each. A linkage-based mapping strategy will utilize either inbred strains or families to search for eQTLs, relying on recombination events between genes and genetic loci with potential regulatory effects. In doing so, there are far fewer markers to analyze; however, this type of study may suffer by missing common variants not present in the studied cohort and may lack the ability to resolve the location of the eQTL to a high degree [1]. Alternatively, using an association-based mapping method necessitates very large sample sizes. This limiting requirement, however, provides greater statistical power, and fine-scale location mapping may be gained as a result of an increased number of recombination events.

In this chapter, we will provide an overview of various publically available bioinformatics software used to conduct an eQTL analysis after one has conducted either microarray or RNA-seq experiments. We being with a worthwhile discussion of preprocessing and normalization steps that ought to be taken when studying expression data, and provide methods to address possible confounders that may be present in an experiment.

2 Confounder Correction and Normalization

Statistical confounders are any unknown variable that may influence a measure of the association between an independent variable and the dependent variable. In an eQTL analysis possible confounders may take the form of population substructure or experimental artifacts. The result of failing to correct for these issues may be an increase in false-positive rate, or spurious association between genetic variants and expression. Normalization addresses concerns of the technical differences in the experimental design and execution among samples. One may therefore apply a normalization procedure to their dataset in order to make their expression values comparable.

2.1 EMMA

EMMA is a useful tool to correct for the effect of population structure on an association mapping analysis [2]. Standing for Efficient Mixed-Model Association, the tool was designed to take into account the genetic makeup of the datasets for eQTL experiments using model organisms. This procedure helps address potential false positives stemming from inherent genetic similarity or familial relatedness within the population being sampled. For example, if half of the dataset is related and the remaining individuals share no relatedness, there would be many implicated genetic loci among the genetically related group which are not true eQTLs and instead they are artifacts of the sampling process. EMMA is able to directly estimate the variance and maximize the likelihood given with fixed-effects integrated out. EMMA performs fast eigendecomposition on a matrix of nonzero eigenvalues and utilizes single-dimensional optimization to efficiently calculate statistical tests. Furthermore, EMMA makes use of a phylogenetic tree as a proxy

Table 1
Tools for data preprocessing and confounder correction

Software	Category	URL
EMMA	Population structure	http://mouse.cs.ucla.edu/emma/
ICE	Population structure	genetics.cs.ucla.edu/ice
AFFYGG	Allelic confounders	github.com/DannyArends/GBIC/tree/master/AffyGG
SVA	Population structure, batch effect	bioconductor.org/packages/release/bioc/html/sva.html
Affy	Normalization	bioconductor.org/packages/release/bioc/html/affy.html
SimpleAffy	Normalization	bioconductor.org/packages/release/bioc/html/simpleaffy.html

for the genetic relatedness between samples in a process called phylogenetic control [3]. EMMA also features built-in computations to apply multiple hypothesis test correction on the maximum likelihood variance component. Altogether, EMMA provides a computationally efficient method to reduce the effect of population substructure when performing an analysis on using inbred strains.

2.2 ICE

The intersample correlation emended (ICE) eQTL mapping method employed by ICE similarly may be used to correct for the heterogeneous makeup of the samples used for an eQTL study. ICE is available for downloaded from the URL given in Table 1, and may also be installed directly as an R package. The need for this type of correction stems from the false-positive associations that arise when there is some degree of invariable regions in the genotypes among a portion of sample studies. ICE is able to incorporate the intersample population structure into a random-effect linear model and correct for this, thereby diminishing the number of inflated p values for identified eQTLs [4].

2.3 SVA

Surrogate variable analysis (SVA) is available as an R package through the bioConductor project. This package can correct for the population structure of the datasets being studied by derived surrogate variables [5]. Surrogate variables are covariates constructed directly from genomic expression data that can be used in subsequent analyses. The SVA package contains functions to iteratively re-weight surrogate variable using a least squares method as a correction procedure for microarray data. For multidimensional RNA-seq data, SVA is also capable of building surrogate variables from the least squares method, but preemptively applies log-transformation to normalize the data by an alternate function. The SVA package provides ComBat as an additional functionality, a method to adjust for batch effect using an empirical Bayes' framework [6]. The ComBat software is useful experimental datasets taken from multiple sources in order to correct for inherent differences in the sets which could misconstrue the results of an eQTL analysis.

2.4 AffyGG

AffyGG is a package that checks for and eliminates deviating probes in Affymetrix microarrays. It has been shown that when an organism has one or more alternate alleles in a given sequence, the probe hybridization may be affected, leading to misleading results in the summary of the probe intensity reading by causing misattributed expression values for genes [7]. This package provides a suite of tools to account for this type of error prior to an eQTL analysis. Similarly, a dependency of AffyGG is the Affy package, which contains functions for exploratory oligonucleotide array analysis. This software can handle various normalization procedures for background correction, such as the widely adapted robust multiarray average, or RMA [8]. SimpleAffy is an available tool that complements the Affy package by providing a stripped-down version of the package. This user-friendly approach may be beneficial to some, as the functions for reading Affy.CEL files and phenotypic data are highly intuitive [9].

2.5 Discussion

Throughout this section we have addressed a handful of software that may be beneficial to avoid spurious associations or false positives. The need to address unknown population structure, potential batch effects, and normalization of the data prior to any statistical procedure are critical to ensure that the conclusions of an eQTL analysis are meaningful.

3 Software for eQTL Analysis

Once the preparatory phase has been completed and any confounding variables have been addressed, a choice must be made as to which software is appropriate given the overall experimental design. In some cases, the study is purely an association-based mapping, while other times one may have a cohort of samples from inbred lines. For either type of study there exists a large collection of computational tools available, and we will attempt to provide an overview of a handful here as a centralized reference. While the collection of tools provided here (Table 2) is in no way comprehensive, it is worth remembering that the open-source software community continually works to build upon existing methods and create unique applications of advanced statistical procedures (all in a variety of programming languages) which means the current list of available software will continue to grow and some may fall out of favor. It may seem fruitless to dwell upon software that may come and go; however, the motivation behind this chapter is to provide a

Table 2
Software for eQTL analysis across various platforms, in order of most recent release

Software	Current release	Release date	Platform	URL
Merlin	1.1.2	12/18/07	Command line	csg-old.sph.umich.edu//abecasis/Merlin/
Pseudomarker	2.04	6/23/08	Matlab	churchill.jax.org/software/archive/pseudomarker.shtml
snpMatrix	2.4	11/11/09	R	bioconductor.org/packages/2.4/bioc/html/snpMatrix.html
eMap	1.2	2/2/10	R	bios.unc.edu/~weisun/software.htm
J/qtl	1.3.4	5/21/13	GUI	churchill.jax.org/software/jqtl.shtml
MapQTL	MapQTL 6	6/4/13	MS - Windows	https://kyazma.nl/index.php/MapQTL/
GridQTL	3.3.0	8/1/13	Browser	gridqtl.org.uk
QTLMap	0.9.7	10/1/13	Command line	forge-dga.jouy.inra.fr/projects/qtlmap
Matrix eQTL	2.1.0	2/24/14	R, others	bios.unc.edu/research/genomic_software/Matrix_eQTL/
PLINK	1.9 beta	5/15/14	Command line	www.cog-genomics.org/plink2
FastQTL	2.184	7/24/15	Command line	fastqtl.sourceforge.net/
R/qtl	1.40.8	10/31/16	R	rqtl.org/

brief overview of commonly used tools in scientific publications, in which eQTL researchers will undoubtedly have an interest.

3.1 Merlin 1.1.2

Merlin is one of the oldest available tools for pedigree-based eQTL analysis; however, it may be used for other types of studies as well. Merlin is capable of identifying nonparametric linkage statistics and can establish observed haplotypes from the data [10]. There is generally a large computational expense required in the representation of all possible pedigree relationships for a given dataset. Merlin reduces this burden through the implementation of sparse binary trees to represent the vectorization of inheritance in a compact way [11]. The trees are then used to establish the likelihoods of the inheritance pattern for single or multiple loci. Merlin has been widely adapted and cited in multiple publications over the years and therefore merits a discussion here, yet it has become less favorable in recent years as there are now more efficient software with lower space requirements that can accommodate modern, large datasets.

3.2 Pseudomarker 2.04

First designed to analyze crosses between inbred lines, Pseudomarker is suite of programs suitable for complex eQTL analysis that has been distributed as MATLAB code. Pseudomarker employs a generalized statistical framework in two steps by first taking into consideration the relationship between the variant with the phenotype, and separately accounts for the location of the variant with respect to the genome. In doing so, a so-called pseudomarker map is then generated to scan the genome for QTLs using Monte Carlo methods for multiple QTL mapping [12].

3.3 snpMatrix 2.4

The R package snpMatrix, available through the bioconductor project, has multiple utilities to test for associations across the entirety of the genome of interest. These applications include confounder correction via principal component analysis to correct for population structure, and imputation. Imputed SNPs are useful to address data in which a small subset of samples possess many polymorphisms that can be typed relative to the much larger number of samples lacking this information. Imputation is viewed as a necessary step prior to performing a meta-analysis, or aggregating data from multiple sources, where the smaller subset becomes the training set upon which the imputed SNPs are derived for samples with missing data. While this package was originally designed for population-based analyses, it was later amended to additionally handle genome-wide association studies within pedigrees [13].

3.4 eMap 1.2

eMap is another R package for eQTL analysis that uses linear regression for cis and trans eQTL mapping based upon a user-defined significance threshold. This linear regression can also be run with permutation or multiple models can be built to take into account both additive and dominant effects simultaneously. The p values are corrected for multiple testing using the false discovery rate (FDR) method. eMap has built in plotting functions to display the results of the eQTL analysis, making the workflow highly streamlined. In addition, eMap can handle identifying and scoring eQTL modules, or regions of the genome that harbor multiple genes and associated eQTLs [14].

3.5 R/QTL

First introduced in 2003, the R package R/QTL made headway in experimental cross mapping of QTLs. Since then, it has been repeatedly extended and improved upon to adapt with advances in genetics and genomics [15, 16]. The current version of R/QTL implements a hidden Markov model algorithm to handle the possibility of missing data in an eQTL analysis. R/QTL allows the user to choose from various methods for genetic mapping. There are built-in functions for quality control checks on the sequence calls, and the software package is capable of searching the sample genome-wide for single- and two-dimensional eQTLs scans. The package is continually in development, and now includes several methods for identifying eQTLs. The repertoire includes interval mapping, Haley-Knott regression, and multiple imputation. One may also fit higher-order QTL models by multiple imputation or the Haley-Knott regression method [17]. The multiple QTL mapping method utilized by R/QTL yields greater power in the resolution of linked QTLs [18]. R/QTL can be extended to handle various covariates as well, and may allow for the mapping of higher-order, multidimensional eQTLs. The authors have also published a companion book, *A Guide to QTL Mapping with R/QTL*. This reference text covers the basics of data input, single-QTL and two-QTL, two-dimensional analyses, experimental design, multiple-QTL modeling, and includes multiple case studies using the package [19]. For those more comfortable with a graphical user interface, the R/QTL package has been rewritten in Java by researchers in The Churchill Group at Jackson laboratory.

3.6 MapQTL 6

For those more adept with MS-Windows, MapQTL version 6 provides a user-friendly and efficient means to perform eQTL mapping. MapQTL is multifunctional software, which equips several mapping methods such as flanking region-based mapping. Not only can it tackle interval mapping, this software can allow for the assumption of normality to be dropped via the Kruskal-Wallis rank sum test. Similar to the latest versions of R/QTL, MapQTL is able to utilize Multiple QTL Mapping, or MQM. The MQM method treats markers as cofactors in order to identify segregating QTLs. The extensibility of the software is also apparent in the testable populations, which include first-generation backcross, recombinant inbred strains, as well as intermated inbred lines among others. The graphical user interface is also capable of generating plots of the resulting analysis [20].

3.7 GridQTL 3.3.0

GridQTL was developed as a means to address growing genomics datasets and built upon previous work of QTLExpress to provide remote access to computing resources [21]. As the name implies, in order to better accommodate user analyses, the web-based portal was built to distribute the workload over a computational grid system. In order to upload a dataset, one must register to gain access to GridQTL; however, the use of GridQTL ensures the storage of analysis output. This shared computational resource thereby facilitates eQTL analysis for researchers without access to sufficient computational resources to map eQTLs via Haley-Knott regression or least squares. Also in place are functionalities to determine linkage disequilibrium and linkage analysis, or LDLA, and efficient methods for haplotyping [22]. As of now, GridQTL remains a free computational resource to researchers investigating eQTLs largely in part by the remaining funding for the project. The same group is looking to address this issue with the transition to

CloudQTL. The project successor will provide cloud-based service for eQTL analysis and tools for a fee paid for by the user [23].

3.8 QTLMap 0.9.7

Similar to the previously mentioned tools, the QTLMap command-line software provides multiple methods to detect eQTLs in data containing unrelated samples dedicated to the detection of QTL from experimental designs in outbred population. With QTLMap, one may perform linkage analysis using linear regression or maximum likelihood. There is also the choice to perform interval mapping with linkage disequilibrium linkage analysis. The software is available to install from source. To reduce computation time spent performing the eQTL analysis, QTLMap has the capability to take advantage of multithreaded parallel processing. This reduction has been taken a step further with the optional provision giving the user the ability to utilize available NVIDIA gpus [24].

3.9 Matrix eQTL 2.1.0

Matrix eQTL has been widely adapted due to its high degree of efficiency in handling the multitude of calculations performed on large-scale datasets. The fast performance is achieved by special preprocessing and expressing the calculation of correlations in terms of large matrix operations [25]. Matrix eQTL supports implementations of multiple models for association testing including least squares additive linear regression or ANOVA for categorical definitions of genotype. A surprising advantage of the implementation of matrix operations is that there are no discernible differences in computation time by including covariates, including for models with correlated and variable error components due to population structure. The p values for cis and trans gene-SNP pairs may then be adjusted with the false discovery rate method in a few additional minutes [26]. Altogether, the popularity of Matrix eQTL is in large part stemming from the reduction in computation time from days to minutes, which facilitates the practicality of testing of multiple models with various covariates considered.

3.10 Second-Generation PLINK: PLINK 1.9 Beta

This represents another example of continual advancement of tools in the eQTL community. Originally published in the American Journal of Human Genetics in 2007, PLINK debuted and was widely accepted for eQTL research [27]. As time went on, PLINK has been succeeded by faster tools with more computationally efficient methods with allowed flexibility in data formatting. More recently, a comprehensive update to the original code and underlying algorithmic framework of PLINK has been released as PLINK 1.9 beta, the testing version of the anticipated PLINK 2 [28]. This new version has improved upon the original methods with dramatic increases in performance when performing Hardy-Weinberg equilibrium for minimal inbreeding populations, and Fisher's exact tests for association testing. For phenotypic prediction, second-generation PLINK provides a lasso method with coordinate descent to allow for penalized regression [29]. As datasets continue to grow in size, this update to PLINK also confers a reduction in the memory requirements, which users without access to large amount of RAM or alternative computing resources will find beneficial. Second-generation PLINK has been developed to seamlessly incorporate data from widely used bioinformatics tools like VCFtools, BCFtools, GATK, and many others [30–32]. In doing so, a new cross-platform genomic relationship matrix is utilized for calculations, which are repeatedly made for a small batch of variants at a time for increased efficiency.

3.11 FastQTL 2.184

FastQTL is one of the newest tools for eQTL analysis and consequent to the dramatic speed increase achievable, the tool has already been adopted by the GTex Consortium for tissue-specific eQTL analysis on their massive datasets [33]. The primary focus of the method for finding candidate *cis*-eQTLs is based upon Pearson correlation, which is also the basis of the method used by Matrix QTL for efficient linear regression. FastQTL runs differ in that they include a direct permutation scheme to repetitively scan the cis-window for the greatest correlation for the variants. This value is then interpreted by its position on a null distribution to assess the corrected significance level for the eQTL. To approximate the adjusted p values, a Beta distribution is used where the shape parameters are given by a maximum likelihood approach. The efficiency of this cis-QTL mapping method is further extended by the ease of use on high performance computing clusters.

4 Concluding Remarks

There are many factors to consider when performing an eQTL analysis; however, the results of such experiments may provide insights toward the regulation of genes at a genome-wide scale. Inferences about the validity and novelty of mapped eQTLs are ultimately made by comparison to prior studies. For human tissues, the GTEx consortium has approached this problem and provides a large reference dataset of genes and eQTLs that may help explain phenotypic variation among differentiated cell types [34]. For instance, the research that produced this dataset found at least one significant eQTL that acts to regulate 31,403 unique genes [35]. Additionally, the Pritchard and Gilad labs provide integrative eQTL datasets and a web portal (*eqtl.uchicago.edu/cgi-bin/gbrowse/eqtl/*) for visualization of alongside other published datasets from a variety of cell types [36]. For nonhuman model organisms, the Jackson Lab hosts a large mouse genomics database (*http://www.informatics.jax.org/mgihome/projects/overview.shtml*) that includes a wealth of eQTL data to aid in research [37].

References

1. Gilad Y, Rifkin SA, Pritchard JK (2008) Revealing the architecture of gene regulation: the promise of eQTL studies. Trends Genet 24 (8):408–415
2. Kang HM et al (2007) Efficient control of population structure in model organism association mapping. Genetics 178:1709–1723
3. Kang HM, Ye C, Eskin E (2008) Accurate discovery of expression quantitative trait loci under confounding from spurious and genuine regulatory hotspots. Genetics 180 (4):1909–1925
4. Leek JT, Johnson WE, Parker HS, Fertig EJ, Jaffe AE, Storey JD (2016) sva: surrogate variable analysis. R Package version 322.0
5. Johnson WE, Rabinovic A, Li C (2007) Adjusting batch effects in microarray expression data using Empirical Bayes methods. Biostatistics 8(1):118–127
6. Alberts R, Vera G, Jansen RC (2008) affyGG: computational protocols for genetical genomics with affymetrix arrays. Bioinformatics 24 (3):433–434. https://doi.org/10.1093/bioinformatics/btm614
7. Chen L, Page GP, Mehta T, Feng R, Cui X (2009) Single nucleotide polymorphisms affect both cis- and trans-eQTLs. Genomics 93:501–508
8. Irizarry H, Collin B-B, Antonellis S, Speed TP (2003) Exploration, normalization, and summaries of high density oligonucleotide array probe level data. Biostatistics 4(1):249–264. https://doi.org/10.1093/biostatistics/4.2.249
9. Miller CJ (2017) Simpleaffy: very simple high level analysis of affymetrix data. http://www.bioconductor.org, http://bioinformatics.picr.man.ac.uk/simpleaffy/
10. Wright FA, Shabalin AA, Rusyn I (2012) Computational tools for discovery and interpretation of expression quantitative trait loci. Pharmacogenomics 13(3):343–352. https://doi.org/10.2217/pgs.11.185
11. Abecasis G, Cherny S, Cookson W, Cardon L (2002) Merlin - rapid analysis of dense genetic maps using sparse gene flow trees. Nat Genet 30:97–101
12. Sen S, Churchill G (2001) A statistical framework for quantitative trait mapping. Genetics 159(1):371–387
13. Clayton D, Leung H-T (2007) An R package for analysis of whole-genome association studies. Hum Hered 64:45–51
14. Sun W (2010) eMap http://www.bios.unc.edu/~weisun/software/
15. Broman KW, Wu H, Sen Ś, Churchill GA (2003) R/qtl: QTL mapping in experimental crosses. Bioinformatics 19:889–890
16. Broman KW (2014) Fourteen years of R/QTL: just barely sustainable. J Open Res Softw 2(1):e11
17. Haley CS, Knott SA (1992) A simple regression method for mapping quantitative trait loci in line crosses using flanking markers. Heredity 69:315–324
18. Arends D, Prins P, Jansen RC, Broman KW (2010) R/qtl: High-throughput multiple QTL mapping. Bioinformatics 26:2990–2992
19. Broman KW, Sen S (2009) A guide to QTL mapping with R/qtl. http://www.rqtl.org/book/rqtlbook_appB.pdf
20. Van Ooijen JW (2009) MapQTL 6, Software for the mapping of quantitative trait loci in experimental populations of diploid species. Kyazma B.V, Wageningen, Netherlands
21. Seaton G, Haley CS, Knott SA, Kearsey M, Visscher PM (2002) QTL Express: mapping quantitative trait loci in simple and complex pedigrees. Bioinformatics 18:339–340
22. Seaton G, Hernandez J, Grunchec JA, White I, Allen J, De Koning DJ, Wei W, Berry D, Haley C, Knott S (2006) GridQTL: A Grid Portal for QTL Mapping of Compute Intensive Datasets. *Proceedings of the 8th World Congress on Genetics Applied to Livestock Production, August 13–18, 2006.* Belo Horizonte, Brazil
23. Allen J, Scott D, Illingworth M, Dobrzelecki B, Virdee D, Thorn S, Knott S (2012) CloudQTL: Evolving a Bioinformatics Application to the Cloud. *Digital Research 2012, September 10–12, 2012.* Oxford, UK
24. Le Roy P, Elsen JM, Gilbert H, Moreno C, Legarra A, Filangi O, INRA (2013) QTLMap https://forge-dga.jouy.inra.fr/projects/qtlmap
25. Shabalin AA (2012) Matrix eQTL: ultra fast eQTL analysis via large matrix operations. Bioinformatics 28(10):1353–1358. https://doi.org/10.1093/bioinformatics/bts163
26. Benjamini Y, Hochberg Y (1995) Controlling the false discovery rate: a practical and powerful approach to multiple testing. J Roy Statistical Society B Meth 57:289–300
27. Purcell S, Neale B, Todd-Brown K, Thomas L, Ferreria M, Bender D et al (2007) PLINK: A tool set for whole-genome association and population-based linkage analysis. Am J Hum Genet 81:559–575

28. Chang et al (2015) Second-generation PLINK: rising to the challenge of larger and richer datasets. GigaScience 4:7
29. Freidman J, Hastie T, Hofling H, Tibshirani R (2007) Pairwise coordinate optimization. Ann Appl Stat 1:302–332
30. Danecek P, Auton A, Abecasis G, Albers CA, Banks E, DePristo MA, Handsaker B, Lunter G, Marth G, Sherry ST, McVean G, Durbin R and 1000 Genomes Project Analysis Group (2011) The variant call format and VCFtools. Bioinformatics 27(15):2156–2158
31. Li H, Handsaker B, Wysoker A, Fennel T, Ruan J, Homer N, 1000 Genome Project Data Processing Subgroup, et al. (2009) The sequence alignment/map format and samtools. Bioinformatics 25:2078–2079
32. McKenna A, Hanna M, Banks E, Sivachenko A, Cibulskis K, Kernytsky A, Garimella K, Altshuler D, Gabriel S, Daly M, DePristo MA (2010) The Genome Analysis Toolkit: a MapReduce framework for analyzing next-generation DNA sequencing data. Genome Res 20:1297–1303
33. Ongen H, Buil A, Brown AA, Dermitzakis ET, Delaneau O (2016) Fast and efficient QTL mapper for thousands of molecular phenotypes. Bioinformatics 32(10):1479–1485. https://doi.org/10.1093/bioinformatics/btv722
34. eQTL. http://eqtl.uchicago.edu/Home.html
35. The GTEx Consortium (2013) The genotype-tissue expression (GTEx) project. Nat Genet 45(6):580–585. https://doi.org/10.1038/ng.2653
36. The GTEx Consortium (2017) Genetic effects on gene expression across human tissues. Nature 550:204–213. https://doi.org/10.1038/nature24277
37. Jackson Lab Mouse Genomics Database: MGI http://www.informatics.jax.org/mgihome/projects/overview.shtml

Chapter 2

Performing QTL and eQTL Analyses with the R-Package GenomicTools

Daniel Fischer

Abstract

We present the R-package GenomicTools that can be used to perform QTL and eQTL analyses in a user-friendly way. First, the theoretical backgrounds of both implemented methods are explained. These are (a) the linear model approach that is commonly used in the standard QTL/eQTL testing as well as (b) a non-parametrical directional testing method implemented in GenomicTools. The directional test overcomes some of the drawbacks of the standard way and is robust against outliers in the data. The main focus, however, is on a detailed explanation, how the R-package is used in practice. Starting from the installation of the package, followed by the data import and also the required steps to perform the analyses, all necessary steps are explained in detail with examples. Also, the commands to create publication-ready figures are presented. The last chapter concludes and discusses general topics related to the analysis of QTL and eQTL data in particular and genomic data in general.

Key words QTL, eQTL, R, Nonparametrics

1 Introduction

The analysis of quantitative traits (QT) reaches back a long time, and structural variants in major genes were already associated with traits over 40 years ago [7]. With the application of modern genomic technologies and methods, it is possible to investigate the individual impact of millions of single nucleotide polymorphisms (SNPs) measurable across the whole genome onto different phenotypes. Countless genome-wide association studies (GWAS) indentified Quantitative Trait Loci (QTL) in practical all domains of life like humans [11], different animal species, e.g. cattle [1], fish [13] or in plants [19]. The results of these QTL studies are typically submitted and organised in large, publicly available databases like, e.g. Animal QTLdb [8] or the Genotype-Tissue Expression (GTEx) project [12].

Primarily, a QTL analysis investigates the association of an SNP with the variation of a measurable feature called *trait*. The most

Xinghua Mindy Shi (ed.), *eQTL Analysis: Methods and Protocols*, Methods in Molecular Biology, vol. 2082,
https://doi.org/10.1007/978-1-0716-0026-9_2, © Springer Science+Business Media, LLC, part of Springer Nature 2020

traits are considered to be polygenic, meaning that more than one gene is affecting the variation and diseases and traits that can be explained by a single SNP are sparse. Examples of polygenic traits are psoriasis, schizophrenia, osteoporosis and, of course, cancer.

Besides the classical QTL analysis, different variations of it have been developed lately. The most prominent is the expression Quantitative Trait Loci (eQTL) analysis [16], where gene expression values are used as quantitative traits. This means, the SNPs are here associated with gene activities. One approach to link an eQTL back to a disease (or any other trait) is to perform first a differentially gene expression analysis to identify genes that have different expression profiles between two trait groups and then perform an eQTL analysis for these genes in order to identify possible SNPs that are affecting the variation of the observed expression profile. Besides QTL and eQTL there are at least 30 different other types of molecular QTL (molQTL) analyses performed, like altered chromatin state (chromQTL), alternative splicing (sQTL), ribosome occupancy (rQTL) or metabolite quantitative trait (mQTL). For an excellent summary please *see* [18].

Almost all commercial and non-commercial genetic data analysis software tools can calculate QTLs and eQTL, e.g. Plink [15]. Also for R there are a few packages that aim for QTL and eQTL analysis. The most prominent and fastest is **MatrixEQTL** [17] for eQTL and **qtl** [2] for QTL analyses. Other packages are e.g. **eqtl** [10], **eQTL** (on Bioconductor), **iBMQ** [9] and **treeQTL** [14]. Here we describe the use of the R-package **GenomicTools** [3] that offers, compared to the software mentioned above, a different statistical method for the association testing. Whereas most solutions rely on linear models or ANOVA type of testing, the here described method uses a non-parametrical test tailored for directional hypotheses as they are present in QTL and eQTL analyses.

2 Background

2.1 The Linear Model Approach

The analysis of QTL and eQTL is mainly based on basic statistics, and the most commonly applied method to correlate an SNP variant with either a phenotype or a gene expression is a linear model. For that, usually, three groups are considered, depending on the observed variants on a specific locus on the genome. In this context, a variant is also called *Allele* and typically two different Alleles (here A and B) are considered, *see*, e.g. Table 1.

In Table 1, e.g. for SNP rs234534, the red coloured Sequence part [A/G] indicates that at the chromosomal position Chr15:97,028,311, there are two different Alleles, respective three different variants, observed, namely A (Allele A) or G (Allele B). In case that either Allele A or Allele B is observed on both

Table 1
Examples of four different SNPs with their flanking sequences in Homo sapiens

SNP	Loci	Sequence	Allele A	Allele B
rs234534	Chr15:97,028,311	CTGAGGA[A/G]GAAAAAT	A	G
rs154325	Chr14:78,988,176	ACTTTGT[C/G]AAGATGT	C	G
rs432234	Chr1:196,827,231	CTCCAAG[A/G]AGATGAT	A	G
rs789345	Chr4:14,496,662	TAAATAA[A/G]AAATTCC	A	G

strands, the variant is called homozygous and is labelled as A/A or B/B, respectively. In case Allele A is present on one strand and Allele B on the other, the variant is called heterozygous and is labelled as A/B. These three different groups are also often marked as 0 (A/A),1 (A/B) and 2 (B/B). These three groups are called here *variant group*.

If we consider now a study population and we are interested in correlating an SNP with either a phenotype (QTL analysis) or a gene expression (eQTL analysis), we assign our study population for each SNP into one of the three corresponding variant groups. Naturally, for each individual within the study also the phenotype or the corresponding gene expression values need to be available. Hence, we create for each individual a two-dimensional dataset, consisting of the variant group and either the phenotype or the gene expression values.

The corresponding mathematical notation would be that we consider a study population of size N. If we measured in total P-many phenotypes, we denote then individual phenotype values as $p_{i,j}$ with $i = 1, 2, \ldots, N$ and $j = 1, 2, \ldots, P$ and for G-many different genes, we write $g_{i,k}$ with $i = 1, 2, \ldots, N$ and $k = 1, 2, \ldots, G$ for the gene expression of individual i and gene k. Individual variant groups are labelled as $v_{i,l}$ with $i = 1, 2, \ldots, N$ and l being the genomic loci of the variant. Possible values of $v_{i,l}$ are then $\{0, 1, 2\}$ depending on the underlying variant group. For example, if $p_{3,2} = 23$ and $g_{1,45} = 1.32$ it means that phenotypes value for phenotype #2 for individual #3 is measured as 23 and for individual #1 the gene expression for gene #45 is measured as 1.32. With that notation we can now write our data in pairs of format $\{(v_{i,l}, p_{i,j})\}$ in case of an QTL study and $\{(v_{i,l}, g_{i,k})\}$ for an eQTL study.

To perform now a single step in a QTL or eQTL study we fix the loci l as well as the phenotype j or gene k and extract all the required information. A few example lines of such a dataset can be found in Table 2 and a visualisation of it in Fig. 1. Here, the x-axis represents the type of variant (indicated as A/A, A/B and B/B and

Table 2
Phenotype data

Individual	Variant	Value
1	A/A	12.3
2	B/B	10.2
3	B/B	9.9
4	A/B	11.3
6	B/B	9.5
7	A/A	14.2
8	A/B	10.2
9	A/B	13.1
10	B/B	14.0
...	...	...

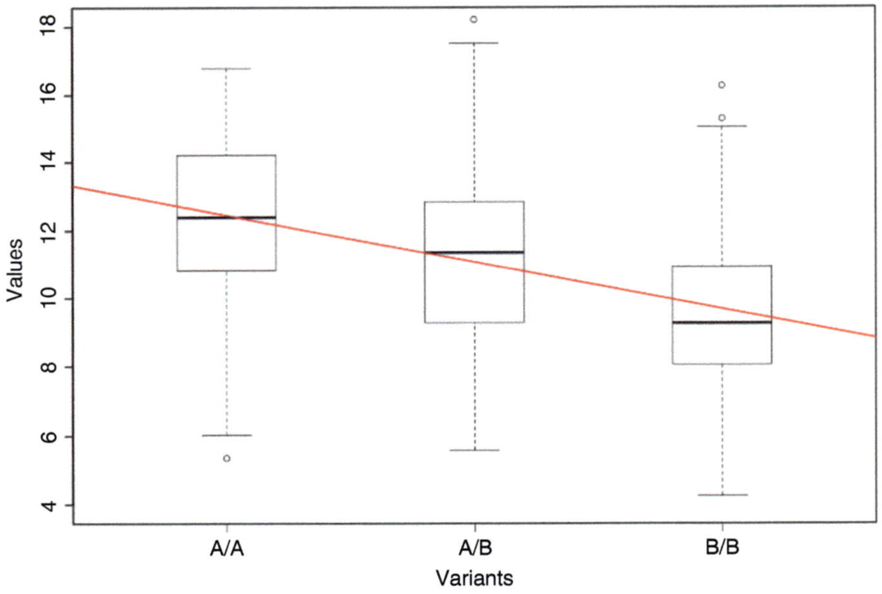

Fig. 1 Genotype vs. phenotype

placed at positions 0, 1 and 2) and the y-axis represents the corresponding value of the gene of interest or the phenotype.

In addition to the raw data, we also added in red a regression line

$$g = av + b$$

where g and v are the gene expression values (g) and variant information (v) from above and a and b are the coefficients to be estimated and which determine the slope (a) and the intercept (b) of the line. For the coefficients a and b usually the Best Linear Unbiased Estimator (BLUE) is searched, and under certain assumptions, this is the Ordinary Least Square (OLS) estimator. In practice, this means that a line is fitted to the data in such a way that the sum of the squared vertical distances between the line and the data points is minimal. The R^2 value gives the goodness-of-fit of this line with respect to the data. In case of a perfect match (all points lay on the line) it is $R^2 = 1$, and 0 if the line has no explanatory power, whatsoever. There is a direct connection between linear models and the Pearson correlation of the data. The goodness-of-fit value R^2 is equal to the squared Pearson correlation between g and v, called $\rho_{g,v}$.

The current standard procedure for (e)QTL testing is based on this linear model approach. To test if there is an association between a variant and a phenotype/gene expression, the following hypotheses are tested:

$$H_0 : a = 0 \qquad \text{vs.} \qquad H_1 : a \neq 0$$

The interpretation of this hypothesis test is simple. In case that $a = 0$ the linear model reassembles a horizontal line, in case that it is not zero ($a \neq 0$), it has either a positive or negative slope. In that case, however, the class of the variant has an impact on the value of the phenotype/gene expression, as the height of the line at A/A is different compared to B/B, whereas if $a = 0$ then the phenotype/gene expression value is similar for A/A and B/B. The above-defined hypothesis is tested with a normal t-test.

Instead of the above-mentioned hypothesis $H_0 : a = 0$ *vs.* $H_1 : a \neq 0$ of the linear model, also the Pearson correlation $\rho_{g,v}$ can be tested with the similar hypothesis $H_0 : \rho_{g,v} = 0$ *vs.* $H_1 : \rho_{g,v} \neq 0$. Here, we test if the correlation between the gene expression/phenotype and the variant is zero or not, indicating whether there is a correlation or not. A t-test for this hypothesis would lead to the same test-statistics as well as the same p-value. In other words, the linear model is just a more intuitive, graphical approach for the same type of test.

2.2 Directional Test

Apparently, the standard (e)QTL testing approach is based on the Pearson correlation, respective the slope of a linear model between the three different variants (as x-values) and the phenotype/gene expressions (as y-values). This approach comes with all the advantages and disadvantages of a parametrical model.

Previously, we proposed a non-parametrical approach [3] to overcome the limitations that originate from using the Pearson correlation for (e)QTL testing based on so-called U-statistics. The underlying theory is explained in detail in [5] and the general tests are implemented in [4]. The method is based on a generalisation of the Mann–Whitney test for directional alternatives of three random variables X, Υ, and Z like

$$H_0 : X = \Upsilon = Z$$

vs.

$$H_1 : X < \Upsilon < Z \qquad \text{or} \qquad H_2 : X > \Upsilon > Z$$

These alternatives are especially suited for the needs in (e)QTL testing, if we assume for H_0, H_1 and H_2 the variable to be $X := A/A$, $\Upsilon = A/B$ and $Z = B/B$. If there is an Allele effect, it would be larger in the homozygous case, compared to the heterozygous case.

If only two variants are observed at a certain locus, the testing problem reduces to

$$H_0 : X = \Upsilon \qquad \text{vs.} \qquad H_1 : X < \Upsilon \qquad \text{or} \qquad H_2 : X > \Upsilon$$

and a standard Mann–Whitney test can be applied. In analogy to the connection between the linear model and the Pearson correlation, we could in that case also test if the Kendall τ correlation between the variant and the phenotype/gene expression equals zero or not as the variant variable is in this case binary. In case of the directional alternative, this is, however, unfortunately not as straight forward.

The directional test approach is based on calculating the probability $P(X = \Upsilon = Z)$ for three random variables X, Υ, Z. In case of the null hypothesis $H_0 : X = \Upsilon = Z$ we would expect this probability equal to $\frac{1}{6}$, as there are in total six different combinations of directional orders possible (e.g. $X < \Upsilon < Z$). If we consider the same example situation as given in Fig. 1 and Table 2, we consider with above notation X_i the phenotype/ gene expression values for individuals that have variant A/A at a particular SNP, Υ_j for those that have variant A/B and Z_k with variant B/B. As a test statistic for $H_1 : X < \Upsilon < Z$, (the test statistic that tests if individuals with the A/A variant at this loci tend to have smaller phenotype/gene expression values than those with variant A/B and these again smaller than those with the B/B variant) we can use then just the sum.

$$\frac{1}{N_X N_\Upsilon N_Z} \sum_i \sum_j \sum_k I(X_i < \Upsilon_j < Z_k).$$

Here is $I(\cdot)$ the indicator function that is 1, if the inner test is true and 0, if not and N_X, N_Υ and N_Z are the group sizes in X, Υ and Z. In other words, we check for all triplet combinations how often the order $X < \Upsilon < Z$ is true. This sum is then either compared to repeated permutation sets of the data to obtain a permutation p-value, which is for large datasets a time-consuming task or p-values are determined based on asymptotics. Asymptotic p-values are much faster to calculate but assume an infinity large sample size. Hence, they are a compromise between speed and accuracy. The default setting in **GenomicTools** is to use asymptotic tests. For details on the asymptotics and the permutation tests, please see again [5] and [4].

This approach avoids the strong parametrical assumptions that are done for the linear model/Pearson correlation approach. For example, there is no distribution assumption for the values within the three variant groups. Further, as this approach is entirely based on ranks, it is very robust against outliers. Besides, the used hypotheses reflect better the assumption of an Allele effect for a variant. Assuming the two Alleles A and B and that Allele B has a particular effect, then the assumption is that the homozygous variant B/B has a stronger effect than the A/B variant. This assumption is exactly modelled in the described test. In case that the A/B variant should have the most significant effect, the here described method could easily be changed to an umbrella hypothesis of the type $H_3 : X < Y > Z$ or $H_4 : X > Y < Z$.

2.3 From Single Tests to (e)QTL Analyses

The methods described in Subheadings 2.1 and 2.2 only consider the tests at a single locus; (e)QTL analyses, however, aim to identify loci along the whole genome (or at least parts of it) that can explain the variability of a particular phenotype and/or a gene expression. Hence, the basis of an (e)QTL analysis is often either a high-density chip that measures up to hundreds-of-thousands variants or is based on whole genome shotgun sequencing ("DNA-seq") [6]. Of course, in exceptional cases are still specific candidate variants tested for their association with a trait. In case of a QTL, usually all available loci are tested against all available phenotypes and the level of association is then typically visualised in a Manhattan plot, where the x-axis represents the loci on the genome and the y-axis represents the level of association, see as an example Fig. 2. The name of the plot originates from the shape of the plot, as it often reminds about the skyline of a city.

In case of an eQTL two genetic features are compared. Whereas in the case of the QTL any measurable feature of interest is linked with genomic markers, we try to identify with an eQTL analysis genetic markers that are associated with gene expression levels. In principle, for an eQTL two different types of approaches are considered, depending on the proximity of the marker to the gene of interest. If genome-wide all markers are tested for association with a gene, this is called *trans*-eQTL, in case of a more regional test (using only markers that are up to 2 megabases (MB) away from the gene of interest) the analysis is called *cis*-eQTL. The typical visualisation for *cis*-eQTL is also a Manhattan-like plot, *see* Fig. 3. However, whereas the typical Manhattan plot is on the log-scale,

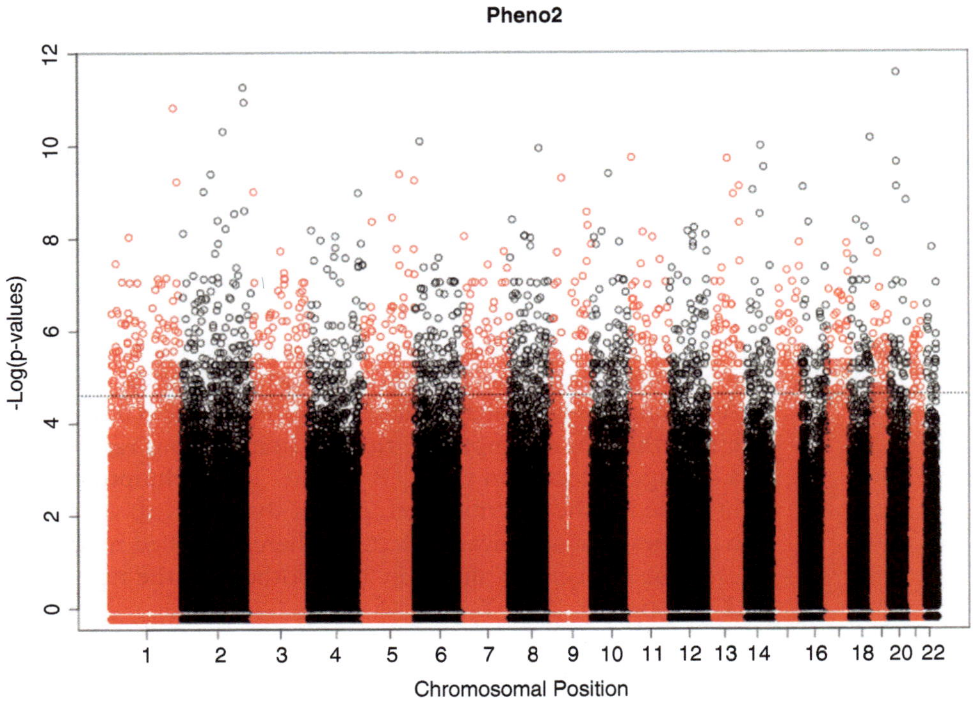

Fig. 2 A Manhattan plot visualises the results of an QTL analysis

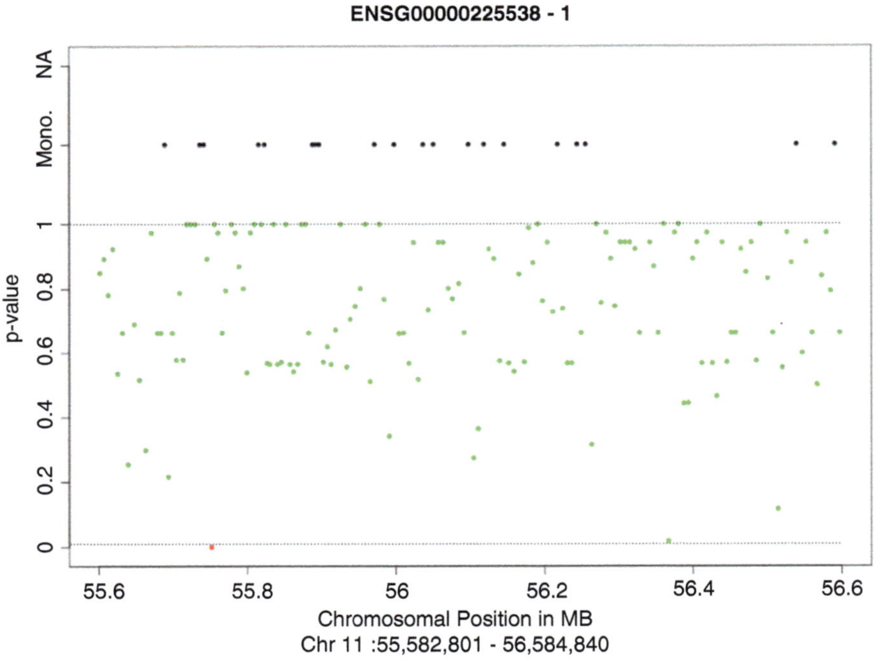

Fig. 3 The visualisation of a *cis*-EQTL result. Missing (NA) and monomorphic (Mono.). SNPs are indicated above the *p*-value area. Non-significant test results are indicated as green dots, red dots are significant test results (for the given level)

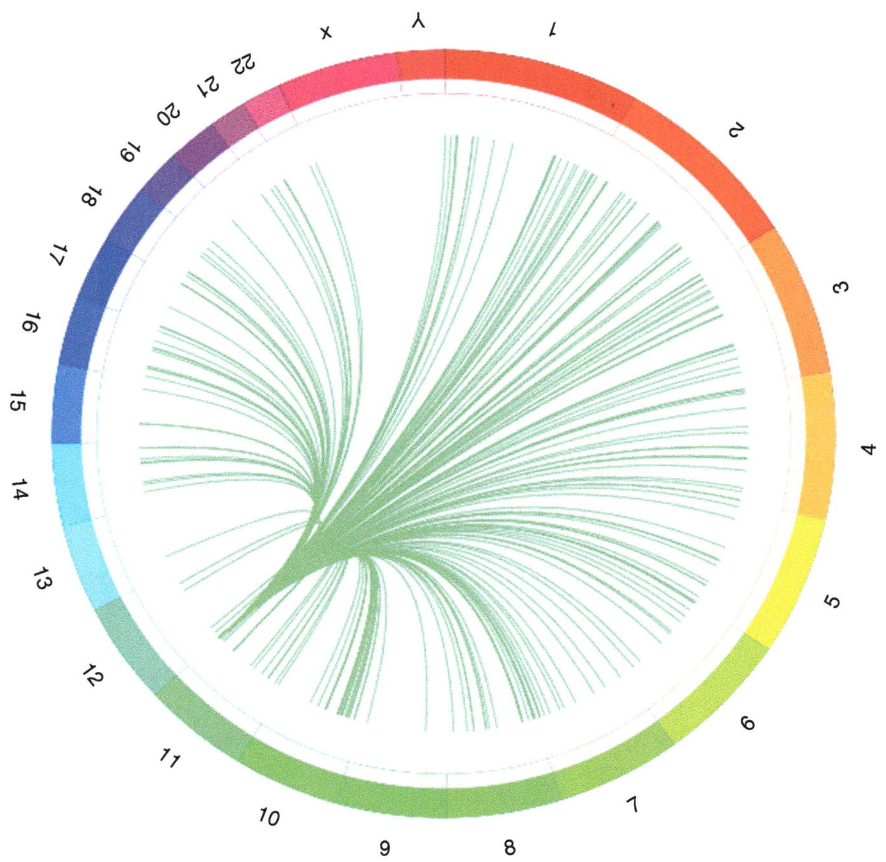

Fig. 4 The visualisation of a *trans*-EQTL

the *cis*-eQTL plot usually shows the real *p*-values. In addition, the homozygous and missing variants are still visualised.

The most common *trans*-eQTL visualisation is done using a circular representation, where a line links gene and marker (*see* Fig. 4) or by using a heatmap-type of figure, where the markers on the genome are placed on the *x*-axis and the gene locations on the *y*-axis. Each significant association is then indicated in the figure with a dot. Naturally, the calculation of (e)QTLs is very computationally demanding, especially for large datasets on the *trans*-scale the amount of performed tests can be in the billions.

2.4 Multiple Testing

When performing QTL and eQTL analyses, there is an issue concerning the total number of performed tests. Assuming that, e.g. for a *trans*-eQTL there are 500,000 SNPs tested against the expression values of a single gene. Here, we receive then also 500,000 *p*-values back from the tests, and if an alpha significance level of 0.001 is applied to extract significant test results, we would already expect in case the null hypothesis holds for all tests just by chance $500,000 * 0.001 = 500$ significant test results. This is statistically a somewhat challenging situation and might require the adjustment for multiple testing. Here, the familiar candidates like the FDR/Benjamini–Hochberg or the Bonferroni correction can be applied. However, as these adjustment methods are maybe too conservative, in practice it might be better to not adjust for multiple testing and identify interesting hotspots and then confirm these with an independent experiment in the lab.

3 Example Workflow

3.1 Preparations for the QTL and eQTL Analyses

In the following section, we describe three typical workflows, one for running a QTL analysis and then one for a *cis-* and a *trans-*eQTL analysis. For that, we apply the R-package **GenomicTools** that includes all the above-described methods. Here, commands that need to be entered by the user or that describe a general syntax are located in boxes with grey background and lines start with `R>`. The corresponding R-output, however, is printed in white boxes.

The installation of the latest stable release version from Cran follows in principle the common R-way. However, the package also depends on the Bioconductor package **snpStats** and requires because of that a few extra steps during the installation. Bioconductor packages are, unfortunately, usually not part of the standard R installation path and consequently, they need to be installed separately. In this case, to prepare for the installation of **GenomicTools** we need to run

```
R> if (!requireNamespace("BiocManager", quietly = TRUE))
+       install.packages("BiocManager")
R> BiocManager::install("snpStats", version = "3.8")
```

Once the Bioconductor requirements are installed, we can proceed the regular Cran way and install **GenomicTools** as any other package located on Cran with

```
R> install.library("GenomicTools")
```

The latest developer version is hosted on GitHub here
https://github.com/fischuu/GenomicTools
and can be installed via the **devtools**:

```
R> library(devtools)
R> install_github("fischuu/GenomicTools")
```

Table 3
Available data and result files

Filename	Description
cisEQTL.RData	*cis*-eQTL result files
transEQTL.RData	*trans*-eQTL result files
phenoQTL.RData	QTL result files
exampleData.csv	Example gene expression data
exampleData.vcf.gz	Example variant data
examplePhenoData.csv	Simulated example phenotype data

The GitHub page of the project is also the preferred way to ask for enhancements, report bugs or ask for general help with the package.

For this example workflow section, we provide a set of example files in a figshare repository that can be found here:

https://figshare.com/projects/Performing_QTL_and_eQTL_analyses_with_the_R-package_GenomicTools/59099

A more user-friendly shortened version of the link is http://bit.ly/2EuCXry.

The figshare project hosts all result files as well as the necessary raw data that can be downloaded to reproduce the example workflows provided here. In detail, Table 3 lists all available data and result files. The result files are provided because some of the calculations run for a while and for the reader, who is interested in reproducing the plots only, these files might be of interest.

The gene expression and variant data originate from the GEUVADIS project [11]. The author is not affiliated to this project but uses here their provided data as an example source. The data files are hosted by the European Bioinformatics Institute (EBI) here:

https://www.ebi.ac.uk/Tools/geuvadis-das/

The files provided via the Figshare repository mentioned above are subsampled and only 1% of the original variant data is provided to keep the computational burden and hardware requirements smaller. Further, the phenotype data is just an example dataset using random numbers.

With the advent of modern workflow and data creation pipelines, the most commonly available file format for genotype data is `vcf`, the variant call format, and this is also here the preferred input file format for the variants. However, **GenomicTools** also provides the option to import a `ped/map` file pair, as provided by Plink. As vcf files are plain ASCII-files and contain plenty of repetitive sequences, they are very well suited for compression. Because of that, **GenomicTools** can also handle compressed vcf files (`*.vcf.`

gz). A typical vcf file can be compressed to up to 10% of its original size, meaning that a vcf.gz file is only 10% of the file size of the underlying vcf file.

All the required file handling is implemented in the subproject **GenomicTools.fileHandler** that is also available on Cran. It is installed as per dependency together with **GenomicTools**.

Before starting an analysis project, it is advisable to create a project folder on the hard drive (HDD) and define the corresponding path variable in R, e.g. called projFolder.

Obviously, the first step is to load the required library

```
R> # Load the required library
R> library(GenomicTools)
```

```
Loading required package: gMWT
Loading required package: clinfun
Loading required package: Rcpp
Loading required package: data.table
data.table 1.12.0  Latest news: r-datatable.com
Loading required package: GenomicTools.fileHandler
```

```
R> # Windows user
R>   projFolder <- "D:\projects\name"
R> # For Linux (example path)
R>   projFolder <- "/home/username/projects/name"
```

This path can then be included in all import/export commands and helps to keep the code readable and easy to change between different operating systems or computers.

To import then a vcf/vcf.gz file, type

```
R> genotypes <- importVCF(file.path(projFolder, "exampleData.vcf.gz"),
+                         na.rm.seq=".")
```

The option na.seq defines here, how missing SNP calls are coded in the vcf file. Typical values are either "." or "./.". Here, we coded them as ".", a single dot.

The gene expression and phenotype data can be imported with the base-R command read.table like this

```
R> # Import the gene expression data
R> expValues <- read.table(file.path(projFolder, "exampleData.csv"),
+                          header = TRUE, sep = " ")

R> # Import the simulated phenotypes
R > phenotypes <- read.table(file.path(projFolder, "examplePheno.csv"),
+                            header = TRUE, sep = "\t")
```

Further, the eQTL analysis requires the gene locations of the genes of interest. For that, we still need to download the gene annotation file that contains the corresponding information. As the example data contains the Ensembl gene IDs, we also download the Ensembl annotation in gtf format from here

ftp://ftp.ensembl.org/pub/release-94/gtf/homo_sapiens/
This folder on the Ensembl server contains the four files

1. `Homo_sapiens.GRCh38.94.abinitio.gtf.gz`
2. `Homo_sapiens.GRCh38.94.chr.gtf.gz`
3. `Homo_sapiens.GRCh38.94.chr_patch_hapl_scaff.gtf.gz`
4. `Homo_sapiens.GRCh38.94.gtf.gz`

from which the regular user usually needs either the `chr.gtf.gz` or the `gtf.gz` file; the `abinitio.gtf.gz` and `chr_patch_hapl_scaff.gtf.gz` are needed for special scenarios. The only difference between the `chr.gtf.gz` and the `gtf.gz` files is how the chromosomes are labelled. In case of `chr.gtf.gz` chromosomes are labelled 'Chr1', 'Chr2', etc. whereas for `gtf.gz` they are just named as '1', '2', etc. The user should download the version that uses the same notation as his or her other files. In our example case, the variants are labelled without leading 'Chr', so the correct file to download would be `Homo_sapiens.GRCh38.94.gtf.gz`.

Once the file is downloaded, it can be imported with the general import function to handle `gtf`-files, that is also included in the **GenomicTools.fileHandler** subpackage. Please note, that this function can also handle compressed files.

```
R> # Import the annotation data
R> annotation <- importGTF(file.path(projFolder,
+                                    "Homo_sapiens.GRCh38.94.gtf.gz"))
```

```
Automatically detected number of rows to skip:  auto
|-------------------------------------------------|
|=================================================|
List of features in column 9:
-------------------------------
gene_id
gene_version
gene_name
gene_source
gene_biotype
```

3.2 Running a QTL Analysis

The function to perform a QTL analysis is `QTL` and takes a couple of arguments, but only two of them are obligatory—the phenotype (`pheno`) and the genotype (`geno`) information. In our case, we have the necessary files already imported, and we are ready to run the QTL analysis. Once the analysis is ready, a couple of standard R S3 functions like `print`, `summary` and `plot` can be applied to the result object.

3.2.1 Performing the QTL

The syntax of the command to run a QTL is

Listing 1 Complete syntax for a QTL analysis:

```
R> QTL(pheno, phenoSamples = NULL, geno = NULL, genoSamples = NULL,
+       method = "dir", mc = 1, sig = NULL, testType = "asymptotic",
+       nper = 2000, which = NULL, verbose = TRUE)
```

but in its most basic form, it is enough to provide the phenotypes and the genotypes

Listing 2 Minimal syntax for a QTL analysis:

```
R> QTL(pheno, geno)
```

The `geno` option expects an object that has previously been imported with the `importVCF` function and the `pheno` object expects a matrix with phenotype data, with one sample per row, and each column being one phenotype. This format is, however, not always provided and sometimes columns and rows are switched so that the phenotypes are in the rows and the individuals are in the columns. In that case, the data matrix needs to be transposed to obtain the correct formatting with the `t( )` -command. In an upcoming version of **GenomicTools** this will be, however, obsolete as the data direction is then estimated directly from the data.

From the optional parameters, the `method` option is important, as it determines the method that should be applied to test for QTLs, namely `LM` or `directional`. The theoretical background of the methods is described in Subheading 2. To match phenotypes with genotypes, the function relies on consistent sample names between the objects. In case they cannot be changed for the original variables, new names for the samples can be given via the two options `phenoSamples` and `genoSamples`. Naturally, the lengths of these two vectors need to match the amount of samples in `pheno` or `geno`. An easy way to speed up the calculation is by setting `mc` to a larger value `1 ( default)` . This option controls the number of used cores and is an easy way to run the calculation parallel. This function, however, uses the forking mechanism and is not supported by Windows machines.

In case we want to perform the QTL just for a subset of the available phenotypes, we can either remove unwanted columns from the `pheno`-object like `pheno[,-c( 1,3,4) ]` (if we want the QTL not to be applied to phenotypes in columns `1`, `3` and `4`) or then specify the name of the phenotype with the `which` option. Assuming that the phenotype object has a column/variable named `weight` and we only want to test QTL for the phenotype, we can add to our `QTL`-command still `which="weight"`.

A somewhat technical option is the `testType`-option. Here we can specify, how *p*-values are computed and options are `asymptotical` or `permutation`. If `permutation` is chosen, the number of permutations can be specified with the `nper`-option. However, be aware that permutation tests are much slower to compute than asymptotical tests and, depending on the number of tested genotypes, this can result in a very long-lasting calculation.

To complete our example workflow, we can perform a QTL analysis limited to the phenotype in row number 2 with the above-imported objects simply by running

```
R> phenoQTL <- QTL(pheno = t(phenotypes[2,]), geno = genotypes)
```

```
We have for 70 % of the samples in the phenotype data the genotype
    information available.
We have for 99.4 % of the samples in the genotype data the phenotype
    values available.
We will test for 1 phenotypes possible QTLs!
________________________________________________
```

Please note, that we also transpose the phenotype data to receive the correct formatting. For further processing, it is advisable to store the output into a variable (here: `phenoQTL`) and also to store the R Object straight away to the HDD. To avoid the file to be overwritten by accident, e.g. if the same script is executed later again, it is practical to add the current timestamp using the `date()` command. The `gsub` function is used to avoid whitespaces in the filename and substitutes those with underscores.

```
R> filename <- paste("phenoQTL\_", gsub(" ", "\_",date()),".RData"
R> save(phenoQTL, file = file.path(projFolder ,filename , sep="")))
```

The above-printed output shows the standard output of the `QTL` function if the option `verbose` is kept as `TRUE`. It shows the user basic statistics of the run, how many matching samples are present and which current phenotype currently is processed.

Once the QTL analysis is finalised, we can either print the results, summarise them or plot them as a Manhattan plot.

3.2.2 Print QTL Results

It is enough just to type the name of the object to print the results of a QTL analysis; in our case this is just `phenoQTL`. However, it is also possible to adjust some of the options to print the output, in that case the `print` command needs to be called explicitly. For example, if all significant associations with p-values smaller than 0.1 shall be printed type

```
R> print(phenoQTL, sig=0.1)
```

```
     Chr          SNP              POS   A B       p-value   Pheno
1270  10  snp_10_7398471   0   7398471   G C  3.289300e-04  Pheno2
2427  10  snp_10_14641610  0  14641610   G A  3.847489e-04  Pheno2
5470  10  snp_10_34853275  0  34853275   A G  2.902426e-04  Pheno2
7933  10  snp_10_58225055  0  58225055   C T  8.298075e-05  Pheno2
```

3.2.3 Summarise QTL Results

There is also an option to summarise the results, using the standard R S3 command `summary`. It will summarise the key facts about the QTL analysis

```
R> summary(phenoQTL)
```

```
QTL Summary
------------------
Type of test          : LM
Tested phenotypes     : 1
```

3.2.4 Plot QTL Results

The most common way to present QTL results is by using the earlier described Manhattan plots. Also, this is implemented as standard S3-class and, if only standard settings are requested, this can be done by typing `plot( phenoQTL)` . In case the default settings are used, it is assumed that the QTL was performed with respect to the human genome and the method takes the genome length information for GRCh37 as provided in Ensembl 68. For other genomes a `data.frame` with chromosome name and lengths can be provided. It is important again that there is a match between the chromosome names provided in this object and the chromosome names used in the QTL analysis. However, in future versions of **GenomicTools** more standard reference genomes will be available.

The corresponding `data.frame` can be created like this

```
R> myGenomeInfo <- data.frame(Chr = c("1", "2", "3"),
+                             Length = c(4345, 1231, 1234))
```

Here the variable `Chr` contains the chromosome names and `Length` the corresponding lengths in bases. For the sake of being concise the chromosome lengths in this example were chosen arbitrary small.

Preparation of a Manhattan plot with a tailored genome information object, one needs to type `plot( phenoQTL, genome = myGenomeInfo)`.

However, in our example case, the default genome is applicable so that we can write

```
R> plot(phenoQTL, log=TRUE)
```

```
Warning message:
In plot.eqtl(phenoEQTL) :
  Warning!!! No genome information provided, use the default (Ensembl
   Human, build 68).
```

We only receive a warning that indicates that the default genome was used, but this we can ignore, as we want to use the default genome.

Another option for plotting a `QTL` object is `log`. The `log` option is a logical flag and indicates if the p-values on the y-axis of the plot should be displayed on a $-\log_{10}$ scale or one the regular [0, 1] interval. However, for regular runs, the default (log-scale) is recommended.

3.3 Running an eQTL Analysis

From the computational perspective, performing an eQTL and a QTL is a very similar process and they mainly differ in the associated quantification object. In a QTL analysis, the numerical values to which genotype groups are associated with are phenotypical information, whereas in the case on an eQTL analysis these are gene expression values. This means, the syntax and options for running an eQTL analysis are very much overlapping with the ones we explained already for the QTL analysis. The main difference between an eQTL and a QTL analysis is that we have in addition to the gene expression values also the gene location available. This additional information leads to slightly other options for the analysis.

The command to run an eQTL analysis is `eQTL`, and the general syntax is of the form

Listing 3 Full syntax for running an eQTL analysis:

```
R>  eQTL(gex=NULL, xAnnot = NULL, xSamples = NULL, geno=NULL,
+          genoSamples = NULL, windowSize = 0.5, method = "directional",
+          mc = 1, sig = NULL, which = NULL, testType = "asymptotic",
+          nper = 2000, verbose = TRUE, IHaveSpace = FALSE)
```

There are a few options that overlap with the `QTL` command, namely `geno`, `genoSamples`, `method`, `mc`, `testType` and `nper` and they are described in Subheading 3.2. From the previous named options, only the `geno` option is required as an input for the function, the other ones are optional.

The three options `gex`, `xAnnot` and `xSamples` handle the gene expression data. The required input `gex` expects the gene expression values with the samples in the rows and different genes in the columns of the matrix. Also, it is possible to test only for a single gene; in that case, the function also accepts a vector as an input. Further, to perform a *cis*-eQTL the genomic location of the tested genes is also required, and the corresponding `gtf`-object (as imported with the `importGTF` function) needs to be given to `xAnnot`. The `xSamples` option has the same purpose as the `phenoSamples` in the QTL function and can be used to specify the sample names of the gene expression data.

To control the local area of a *cis*-eQTL, one can use the `windowSize` option. Here, the *cis* area in megabases (MB) can be specified. Please notice that the effective window is twice as large as the window size, as the option is applied down- and upstream of the gene. Further, the window size differs between different genes, as the window is not given with respect to the centre of a gene, but to the start and end of the corresponding flanking exons. The `windowSize` is also used to switch between a *cis*- and *trans*-eQTL analysis. In case that a value is given to this option, a *cis*-eQTL is performed. If the `windowSize` is set, however, to `NULL`, meaning that the `windowSize` is empty, then all available SNPs are tested against all available genes. In other words, a *trans*-eQTL is initiated.

In our running example, the most simple calls to perform a *cis*-eQTL and a *trans*-eQTL are

```
# Perform a cis-eQTL
R> cisEQTL <- eQTL(gex = t(expValues),
+                  xAnnot = annotation,
+                  geno = genotype)
```

```
xAnnot is given in gtf format (from importGTF). We transform it with gtfToBed() into required bed-
    format.
We will transform the gene annotations into a list ... done (Fri Feb  1 08:08:11 2019)!
We have for 70 % of the samples in the expression data the genotype information available.
We have for 99.4 % of the samples in the genotype data the expression values available.
We have for 75 % of the expression data the annotations.
We will test for 1000 genes possible eQTLs!
---------------------------------------------
We calculated eQTLs for ENSG00000225538 for 163 SNPs (Fri Feb  1 08:08:17 2019)
We calculated eQTLs for ENSG00000237851 for 136 SNPs (Fri Feb  1 08:08:20 2019)
(...)
```

```
# Perform a trans-eQTL
R> transEQTL    <- eQTL(gex = t(expValues),
+                       xAnnot = annotation,
+                       geno= genotypes,
+                       windowSize = NULL,
+                       which=c("ENSG00000151503"))
```

```
xAnnot is given in gtf format (from importGTF). We transform it with gtfToBed() into required bed-
    format.
We will transform the gene annotations into a list ... done (Fri Feb  1 08:09:48 2019)!
We have for 70 % of the samples in the expression data the genotype information available.
We have for 99.4 % of the samples in the genotype data the expression values available.
We have for 100 % of the expression data the annotations.
We will test for 1 genes possible eQTLs!
---------------------------------------------
We calculated eQTLs for ENSG00000151503 for 381,865 SNPs (Fri Feb  1 10:47:05 2019)
```

Please notice that we applied again the transpose function `t()` to the `expValues` object to bring the expression matrix into the correct direction (samples in the rows, genes in the columns). In this example, the *trans*-eQTL needs still some further explanations. The first three options do not differ between the two commands, and the magic happens in the following two lines. As indicated above, by setting the windows size to `NULL` instead of a numerical value, we initiate a *trans*-eQTL. Using the `which` option restricts our tested genes. The gene expression object contains the expression values for 1000 genes, and on a current workstation, the calculation of a *trans*-eQTL for 500,000 SNPs takes about 10 h calculation time when only one calculation core is used. That means the calculation of all available *trans*-eQTL requires the use of a computation cluster to run this task effectively. The syntax how a large *trans*-eQTL is split to a cluster goes over the scope of this example workflow, a functional script set for a computer cluster using the slurm queueing system can be found on the project webpage http://genomictools.danielfischer.name.

By default, the `eQTL` function in *trans*-mode stores only significant test results. However, if the user wishes to keep all test results, the option to set the significance level of results to be stored needs is `sig=1`. However, this results usually in huge data objects with hardware requirements that exceed the standard workstation's specifications. For that reason, another security flag needs to be set, to ensure the user knows truly what he or she does. Hence, for running that type of *trans*-eQTL the user needs still to set the logical flag `IHaveSpace=TRUE`.

3.3.1 Print eQTL Results

Similar to the QTL results, an eQTL result can be printed on the screen using the standard R S3 class command `print`. This means again that for the standard options, it is enough to type the object name into the console to print the content of the object and the `print` function only needs to be used for specifying additional options. The standard output for a *cis*- and `trans`-eQTL looks like this

```
R> cisEQTL
```

```
  Chr     Start       End              Name            Gene      p.value
2  11  55734322  55734322  snp_11_55734322 ENSG00000225538 2.179503e-11
3   6 142924206 142924206 snp_6_142924206 ENSG00000237851 1.019492e-04
4   6 143156710 143156710 snp_6_143156710 ENSG00000237851 5.163365e-03
5   6 143180433 143180433 snp_6_143180433 ENSG00000237851 3.589815e-03
```

```
R> transEQTL
```

```
   chr                  SNP Location      p.value      Assoc.Gene
1:  10       snp_10_9993635  9993635 0.0008437778 ENSG00000151503
2:  10      snp_10_10792136 10792136 0.0008461591 ENSG00000151503
3:  10 indel:1I_10_19139958 19139958 0.0002820548 ENSG00000151503
```

3.3.2 Summarise eQTL Results

The `eQTL` function also carries a `summary` function that summarises the key facts of an eQTL run concisely. The summary output after typing `summary(cisEQTL)` and `summary(transEQTL)` looks like

```
R> summary(cisEQTL)
```

```
cis-EQTL Summary
----------------
Type of test                                          : directional
Tested genes                                          : 876
Total number of SNPs (in geno object)                 : 381865
Window size (in MB)                                   : 0.5
Average (median) number of SNP in window              : 134.7352 ( 142 )
Average (median) number of sig. (p< 0.01 ) SNP in window : 2.474886 ( 1 )
```

```
R> summary(transEQTL)
```

```
trans-EQTL Summary
______________
Type of test                              : directional
Tested genes                              : 1
Total number of SNPs (in geno object)     : 381865
Window size (in MB)                       : trans-eQTL
```

3.3.3 Plot eQTL Results

The eQTL analysis results can be visualised with corresponding plots, in case of a *cis*-eQTL, this is a scatterplot that indicates for each SNP within the tested window the corresponding *p*-value, if the SNP contained only missing data or if the SNP is homozygous. Further, significant test results can be highlighted and the level of significance can be adjusted with the `sig` option. When performing a *cis*-eQTL usually many genes are tested so that they cannot be displayed all at once on the screen. To overcome this, two options are available. Either the user can specify the `file` option and store all figures individually on the HDD, or to use the `which` option and specify a particular gene to display the result in a graphic window.

```
# Plot the results for the first gene
R> plot(cisEQTL, which=1)
# Plot the results for the gene names ENSG000123
R> plot(cisEQTL, which="ENSG000123")
# Store all the plots in the folder /home/user/figures
R> plot(cisEQTL, file="/home/user/figures")
```

While the figures of *cis*-eQTL analyses usually contain all test results for a specific association, this is not feasible for a *trans*-eQTL. **GenomicTools** offers for this a circular plot to indicate genome-wide associations of an SNP with a gene expression. Here, chromosomes are indicated by colours and arches connect the gene location with the associated SNP. The colour of the connection is similar to the colour of the chromosome on which the gene is located.

To visualise a *trans*-eQTL, the object needs to be plugged into the `plot` function

```
# Plot a trans-eQTL result
R> plot(transEQTL)
```

4 Notes

Performing association studies leads usually always to significant results and the interpretation of the results comes with great responsibility and the acceptance of significant results and their interpretation turns also out to be difficult. When interesting candidate regions are hunted, it is often desired to have a block of several SNPs being associated at once with a specific trait. The need for this becomes especially then apparent, when we have a look at the here presented phenotype data, which is entirely sampled at random and should not contain any significant results. However, the way we sampled the random data from a normal distribution also allows for rare, but extreme values and they appear to be significant test results, although they follow the same distribution as any other data point. This should serve as a cautionary note to be careful, even (or especially with) significant results in (e)QTL studies. However, on the other hand, especially for diseases exist causative mutations, where a single SNP is responsible for the presence of a disease and in this case the results of single SNPs are in fact of utter importance and significance.

Another more general note for dealing with genomic data is the advice to keep data as often as possible compressed. The compression rate is usually very high for this kind of data, and there is excellent potential to save disc space and resources. Modern computer systems with large file storages make it easy to be sloppy with disc space and especially here is a significant potential in saving money. Backup plans are usually paid per used gigabyte (Gb) disc space and by compressing files, this cost can be reduced a lot.

One point that we have not considered above is the Minor Allele Frequency. Usually, it does not make sense to calculate (e) QTL for Alleles that have an MAF of 5% or less, so it would be advisable to filter out those loci prior to the analysis. This is also a planned feature for future versions of **GenomicTools**.

Finally, it is worth to mention that **GenomicTools** is an actively developed tool and if a particular feature is requested, it is almost always a good idea to ask the author, if it would be possible to implement it. Also, any reports on bugs are always highly welcomed.

References

1. Bouwman A, Daetwyler H, Chamberlain A, Ponce C, Sargolzaei M, *et al* (2018) Meta-analysis of genome-wide association studies for cattle stature identifies common genes that regulate body size in mammals. Nat Genet 50 (3):362–367. https://doi.org/10.1038/s41588-018-0056-5
2. Broman K, Wu H, Sen S, Churchill G (2003) R/qtl: QTL mapping in experimental crosses. Bioinformatics 19:889–890. https://doi.org/10.1093/bioinformatics/btg112
3. Fischer D (2017) The r-package genomic tools for multifactor dimensionality reduction and the analysis of (exploratory) quantitative trait loci. Comput Methods Prog Biomed 151:171–177. https://doi.org/10.1016/j.cmpb.2017.08.012
4. Fischer D, Oja H (2015) Mann-Whitney type tests for microarray experiments: TheRPackagegMWT. J Stat Softw 65(9). https://doi.org/10.18637/jss.v065.i09
5. Fischer D, Oja H, Schleutker J, Sen P, Wahlfors T (2013) Generalized Mann-Whitney type tests for microarray experiments. Scand J Stat 41(3):672–692. https://doi.org/10.1111/sjos.12055
6. Fleischmann R, Adams M, White O, Clayton R, Kirkness E, *et al* (1995) Whole-genome random sequencing and assembly of Haemophilus influenzae Rd. Science 269 (5223):496–512. https://doi.org/10.1126/science.7542800
7. Gelderman H (1975) Investigation on inheritance of quantitative characters in animals by gene markers. Theor Appl Gen 46:300–319
8. Hu ZL, Park C, Reecy J (2018) Building a livestock genetic and genomic information knowledgebase through integrative developments of animal QTLdb and CorrDB. Nucleic Acids Res 47(D1):D701–D710. https://doi.org/10.1093/nar/gky1084
9. Imholte G, Scott-Boyer MP, Labbe A, Deschepper C, Gottardo R (2013) iBMQ: a r/bioconductor package for integrated Bayesian modeling of eQTL data. Bioinformatics 29 (21):2797–2798. https://doi.org/10.1093/bioinformatics/btt485
10. Khalili A, Loudet O (2012) eqtl: tools for analyzing eQTL experiments: a complementary to Karl Broman's 'qtl' package for genome-wide analysis. R package version 1.1-7. https://CRAN.R-project.org/package=eqtl
11. Lappalainen T, Sammeth M, Friedländer M, 't Hoen PC, Monlong J, *et al* (2013) Transcriptome and genome sequencing uncovers functional variation in humans. Nature 501 (7468):506–511. https://doi.org/10.1038/nature12531
12. Lonsdale J, Thomas J, Salvatore M, Phillips R, Lo E, *et al* (2013) The genotype-tissue expression (GTEx) project. Nat Genet 45 (6):580–585. https://doi.org/10.1038/ng.2653
13. Lv W, Zheng X, Kuang Y, Cao D, Yan Y, Sun X (2016) QTL variations for growth-related traits in eight distinct families of common carp (Cyprinus carpio). BMC Genet 17(1). https://doi.org/10.1186/s12863-016-0370-9
14. Peterson C, Bogomolov M, Benjamini Y, Sabatti C (2016) TreeQTL: hierarchical error control for eQTL findings. Bioinformatics 32 (16):2556–2558. https://doi.org/10.1093/bioinformatics/btw198
15. Purcell S, Neale B, Todd-Brown K, Thomas L, Ferreira M, *et al* (2007) PLINK: a tool set for whole-genome association and population-based linkage analyses. Am J Hum Genet 81 (3):559–575. https://doi.org/10.1086/519795
16. Rockman M, Kruglyak L (2006) Genetics of global gene expression. Nat Rev Genet 7 (11):862–872. https://doi.org/10.1038/nrg1964
17. Shabalin A (2012) Matrix eQTL: ultra fast eQTL analysis via large matrix operations. Bioinformatics 28(10):1353–1358. https://doi.org/10.1093/bioinformatics/bts163
18. Vandiedonck C (2018) Genetic association of molecular traits: a help to identify causative variants in complex diseases. Clin Genet 93 (3):520–532. https://doi.org/10.1111/cge.13187
19. Zhang X, Huang C, Wu D, Qiao F, Li W, *et al* (2017) High-throughput phenotyping and QTL mapping reveals the genetic architecture of maize plant growth. Plant Physiol 173 (3):1554–1564. https://doi.org/10.1104/pp.16.01516

Chapter 3

eQTL Mapping Using Transcription Factor Affinity

Elisa Mariella, Elena Grassi, and Paolo Provero

Abstract

In the last decades, thousands of common genetic variants have been associated with human diseases by genome-wide association studies (GWAS). However, the functional interpretation of GWAS hits is usually nontrivial, especially because most of them lay outside the coding genome. These noncoding variants presumably exert their effect by altering gene expression levels; therefore, expression quantitative trait loci (eQTL) mapping analyses represent an important step in understanding their functional relevance and identifying the target genes. Here we describe an alternative strategy for the detection of eQTL that takes into account the combined effect of genetic variants within regulatory regions and leverages the idea that changes in gene expression often are the consequence of the alteration of transcription factor (TF) binding.

Key words Human genetic variants, GWAS, eQTL, Transcription factors, Total binding affinity

1 Introduction

Understanding the effect of genetic variants on human diseases has always been a major goal in the biomedical field, and in the past, several rare alleles causing Mendelian diseases have been identified through linkage-based family studies. This field of research has been recently revolutionized thanks to progress in sequencing technologies occurred after the publication of the first draft of the human genome [1, 2]. Indeed, in the last decades, the availability of increasingly large catalogs of human genetic variants [3] and the advent of genome-wide association studies (GWAS) [4] have uncovered thousands of unexpected associations between common genetic variants and complex human diseases. Nevertheless, the functional interpretation of these GWAS hits remains problematic, especially because the majority of them are located within the noncoding portion of the human genome [5–7].

Useful information can be derived from the identification of genetic variants affecting quantitative molecular phenotypes that can act as functional mediators between a genetic variant and a disease of interest. For example, expression quantitative trait loci

Xinghua Mindy Shi (ed.), *eQTL Analysis: Methods and Protocols*, Methods in Molecular Biology, vol. 2082,
https://doi.org/10.1007/978-1-0716-0026-9_3,

(eQTL) mapping analyses look for statistical correlations between single genetic variants and gene expression levels [8]. Although this approach has undeniably given substantial contributions to the functional interpretation of the genetic variants associated to disease, it has two main limitations: it disregards our incomplete but substantial knowledge of the regulatory code, and the combined effect of several variants on the expression of the same gene cannot be easily studied. In a more sophisticated model, we can assume that all the genetic variants within a regulatory region contribute to determine the binding affinity of transcription factors (TFs), which in turn results in a change of the expression of a target gene.

Positional weight matrices (PWMs) [9] are commonly used to represent the binding preferences of TFs as the frequency of each nucleotide in each position within the binding site. In order to summarize the effects of multiple genetic variant within regulatory regions, the PWMs of many TFs can be used to compute the total binding affinity (TBA) [10] of the regulatory sequences obtained in several individuals. With a single number, the TBA measures the likelihood that a TF described by a PWM will bind a particular DNA sequence, taking into account the contribution of both high and low affinity binding sites. Here we are going to describe an alternative approach for the detection of eQTLs in which the TBA profile of a regulatory region is correlated with the expression levels of a target gene through principal component regression (PCR) [11], in order to identify genes whose expression variation across a large human cohort is driven by noncoding genetic variants acting through the alteration of the binding of TFs. Figure 1 depicts a global view of the pipeline.

2 Materials

This section provides a list of the software packages needed for analysis.

1. Some functions from EIGENSTRAT [12] allow performing a principal component analysis (PCA) on genotypic data to correct for the population structure [13].
2. The Rust library vcf-rider is used to reconstruct the sequence of the regulatory regions of the individuals, and to compute the TBA values on all these sequences, in a computationally efficient way (*see* **Note 1**).
3. Different functions from both vcf-tools [14] and PLINK [15] are used for the manipulation of VCF files.
4. The fitting of TBA models through PCR is performed using several R functions. In the supplementary materials (https://gitlab.com/ProveroLab/tba-eqtl) an example of an R script to fit these models is available, along with an R script that generates the TBA matrix (see below) starting from the output of vcf-rider.

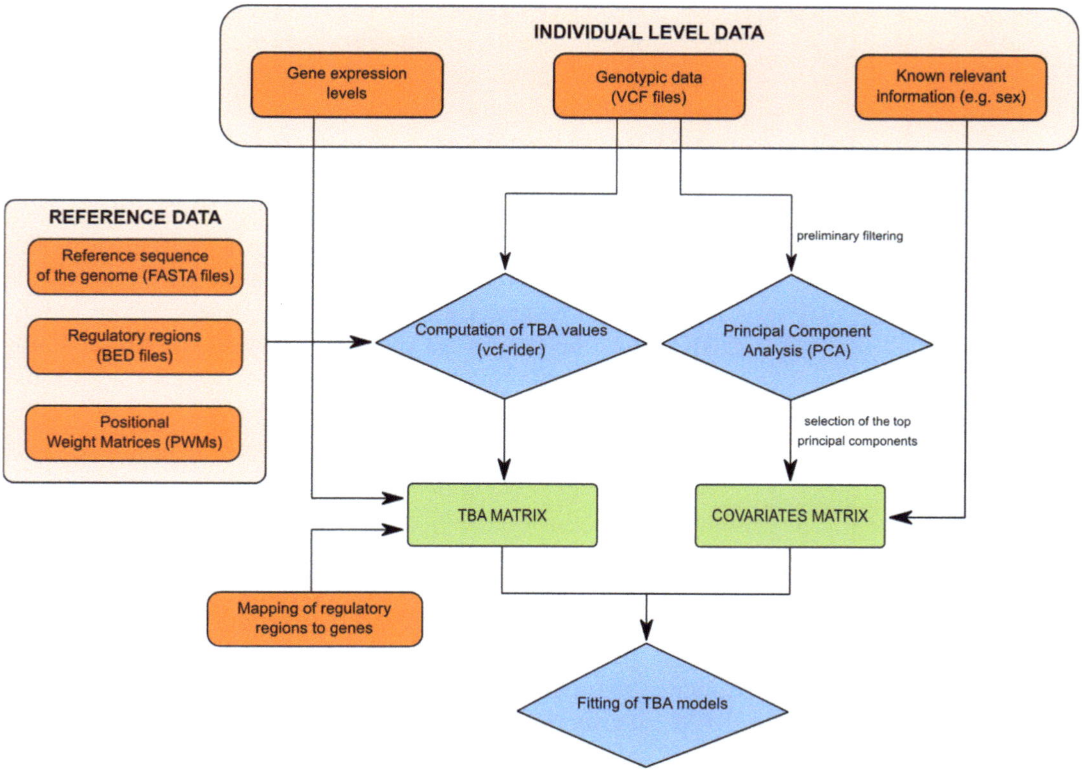

Fig. 1 Schematic representation of the described method for the eQTL detection. The implementation of the procedure requires a large dataset, in which genotypic data and gene expression levels are available for many individuals. In addition, the reference sequence of the genome for the species of interest, the coordinates of some regulatory regions and a collection of positional weight matrices (PWMs) are needed. In the first step, the top principal components derived from genotypic data are combined with other relevant information to generate the matrix of covariates. Then, starting from the reference sequence of the genome, genotypic data are used to reconstruct the sequence of the regulatory regions in all the individuals, and to compute at the same time, for each individual, the total binding affinities (TBA) for a set of transcription factors whose binding preferences are described by positional weight matrices (PWMs). Both these steps can be performed in a computationally efficient way using the Rust library vcf-rider. Finally, the TBA values of each regulatory region are combined with the expression levels of its target gene to produce the TBA matrix that is then used, together with the matrix of covariates, to fit the TBA models. In this way, genes whose expression variation across individuals is driven by regulatory genetic variants that affect the binding of transcription factors can be identified

3 Methods

3.1 Obtain All the Input Files

Before starting the analysis, make sure you have all the input files that are described in this section. *See* **Note 2** for additional details.

1. Download FASTA files with the reference sequence of the genome for the species of interest (reference.fa) from a genomic database (such as Ensembl and UCSC genome browser).

2. Get a collection of regulatory regions (regulatory.bed) and a file mapping them to the corresponding target genes (mapping.tsv). The coordinates of the regulatory regions must be specified using the standard BED format, and they must be sorted with respect to the start coordinate. *See* **Note 3** for additional details.
3. Obtain a matrix with RNA-Seq derived expression levels for a set of individuals (expression.tsv). Columns must map to the different individuals in the study, while rows refer to the genes. In addition, a header containing the identifiers of the individuals should be present as the first row, and the first column should contain the gene identifiers. *See* **Note 4** for additional information about gene expression data.
4. For the same individuals, obtain VCF files including phased genotypes from WGS data (wgs.vcf). *See* **Note 5** for additional details.
5. Download the PWMs representing the binding preferences of a set of TFs (pwms.tsv). Different sources are available, for example, the JASPAR [16] and HOCOMOCO [17] databases include PWMs for several TFs in multiple species. All the downloaded PWMs must be listed in a headerless file with PWM identifiers on the first column, PWM positions in the second column (this column is ignored by vcf_rider), followed by 4 columns with A/C/G/T pseudo-counts (i.e., integer > 0). *See* **Note 6** for additional details.
6. Get the background frequencies of A/C/T/G nucleotides in the set of sequences of interest (e.g., in [11] the human intergenic regions were used), needed to compute TBA scores. These values must the reported in a single tab-delimited line (bg.tsv).

3.2 Preparation of the Matrix of Covariates

1. Starting from the original VCF files, perform a LD-based pruning of the genetic variants exploiting the --indep-pairwise function of PLINK with the following recommended parameters: window size = 50 SNPs, step size = 5 SNPs, and r^2 threshold = 0.8.
2. Combine the --exclude-bed, --maf, --snps, and --recode functions from vcf-tools in order to generate new VCF files resulting from the following transformations:
   (a) Exclusion of long-range LD regions.
   (b) Selection of common genetic variants (i.e., those having minor allele frequency > 0.05).
   (c) Selection of the genetic variants identified by the LD-based pruning.

3. Perform a principal component analysis (PCA) on the resulting genotypic data exploiting the smartpca.perl script from EIGENSTRAT.
4. Generate a single matrix (covariates.tsv) including the top principal components of the genotypic data and any other relevant covariate (e.g., sex) (*see* **Note 7**).

3.3 Preparation of the TBA Matrix

1. To obtain TBA values for a set of regulatory regions use:

   $ vcf_rider -v wgs.vcf -p pwms.tsv -b regulatory.bed -r reference.fa -f bg.tsv

   [-a regulatory_snps.tsv] > tba.tsv

   The main output (tba.tsv) is a headerless file with the following fields: bed identifier, start coordinate, end coordinate, PWM identifier, individual identifier, allele, and TBA value. Please note that TBA values for the two alleles are reported separately here, but in the rest of the pipeline we refer to the TBA as the sum of the two allele values. In addition, when the optional -a parameter is used, the tool also produces a file (regulatory_snps.tsv) reporting the list of genetic variants found in each regulatory region (*see* **Note 8**).
2. Generate the TBA matrix containing all the dependent and independent variables used for the fitting of TBA models.
   (a) For each regulatory region and each individual, sum the TBA values that have been independently computed for the two alleles.
   (b) Compute the $\log_2$ of all the resulting TBA values.
   (c) We strongly recommend eliminating those regulatory regions in which the presence of structural variants or large indels causes an excessive variation in the sequence length among the individuals, so as to avoid differences in the TBA values strongly driven by differences in regulatory sequence length. For example, in [11] we discarded all the regulatory regions that, at least in some individuals, differed by more than 10% in length from the reference sequence (*see* **Notes 8** and **9**).
   (d) Produce a matrix in which each column refers to one PWM and each row contains all the TBA values obtained from the regulatory region of a single individual.
   (e) Add an extra column to the TBA matrix, associating to each regulatory region the expression levels of the target gene (*see* **Note 10**). The first column should contain an identifier composed of the id of the individual, the id of the gene, and the id of the regulatory region.

3.4 Fitting of the TBA Models

1. Load both the TBA matrix and the covariates matrix in R as data frames.
2. Fit an independent TBA model for each pair composed by a regulatory region and one of its target genes performing PCR (*see* **Note 11**). This task can be efficiently implemented exploiting the ddply function from the CRAN R package plyr [18] to apply the same function to each block of the TBA matrix. In particular, for the fitting of each model the following operations must be performed:
   (a) Compute the PCs of the TBA values using the prcomp function and then select enough principal components to explain a preset percentage of the variance of the TBA values (e.g., those that cumulatively explain at least the 95% of the variance).
   (b) Use the lm function to fit two nested linear models:
      - In the outer model, both the selected TBA-derived principal components and all the covariates are used as predictors to regress the gene expression values.
      - In the inner model, only the covariates are used as predictors to perform the same regression.
   (c) Exploit the anova function to perform an F-test comparing the two nested models and obtain the resulting *p* value.
   (d) Get an empirical *p* value by repeating the fitting of the models multiple times after the random shuffling of gene expression values across individuals.
3. Correct the empirical *p* values for multiple testing applying the Benjamini-Hochberg approach to get adjusted *p* values, and then select the significant TBA models comparing each corrected empirical *p* value with a threshold (e.g., requiring the adjusted *p* value to be <0.05) (*see* **Note 12**).

4 Notes

1. Reconstructing the whole genome of individuals from VCF files is a trivial task, but performing computations on them, such as counting the number of binding sites for a given transcription factor, can become time consuming for a large number of individuals. This task can clearly be implemented in a more efficient way by factoring out the non-polymorphic portions of the genome, and considering them only once for all individuals. Also polymorphic portions can be evaluated only once for each haplotype, rather than for each individual. To efficiently compute the TBA scores for different individuals

we developed a library to compute generic sequence-based scores on a set of nonoverlapping genomic intervals, starting from a VCF file and the reference sequence. The library is written in Rust, a modern compiled language similar to C/C ++ and providing an easier programming environment in terms of type management and memory safety, while nonetheless achieving comparable performance. It is a generic tool that can be easily adapted for the computation of any sequence-based score. Detailed instructions about its installation can be found at https://github.com/vodkatad/vcf_rider.

2. Independent FASTA, VCF, and BED files must be prepared in case of multiple chromosomes, because the processing of one chromosome at a time is mandatory for the computation of TBA values. In addition, all of them must, of course, refer to the same version of the selected genome (e.g., the hg38 version for the human genome). A collection of example files is provided as supplementary material, along with R scripts for the TBA matrix generation and model fitting.
3. Both local and distal regulatory regions can be taken into account. In the simplest approach, local regulatory regions can be defined based on gene coordinates, while additional data, such as histone modification profiles or chromatin conformation data, are necessary to define cell-type-specific distal regulatory regions and their association to genes. For example, in [11] we defined promoters as the regions spanning 1500 bp upstream and 500 bp downstream from each transcription start site (TSS), while a collection of cell-type-specific enhancers, together with the associated genes, was obtained using the PreSTIGE algorithm [19].
4. The measured gene expression levels can be affected by different types of technical artifacts: their preliminary cleaning, for example, in order to remove a batch effect, usually results in an increased number of detected associations in eQTL mapping analyses and therefore is strongly recommended. For example, this goal can be achieved using probabilistic estimation of expression residuals (PEER) [20], a collection of Bayesian approaches able to detect a small number of hidden factors explaining much of the variance, and then factor them out from gene expression data. The TBA model was initially developed exploiting a dataset in which RNA-Seq data were available; however, also microarray-derived gene expression values can be used.
5. In VCF files the genotypes are phased when the alleles of all the genetic variants have been attributed to the maternal or paternal chromosome in a consistent way [21]. Phasing is mandatory for the implementation of this analysis, because otherwise the

sequences of the two alleles of each regulatory region cannot be obtained. In addition, the successful reconstruction of these sequences requires the knowledge of all the genetic variants of each individual, which however is not provided by SNP arrays. This could be an important limitation, because genotyping with SNP arrays is still more common than WGS in large human cohorts. However, fortunately, the problem can be overcome by performing genotype imputation, i.e., predicting the genotypes not directly assayed using a reference panel [22]. Several tools have been developed for both phasing and genotype imputation, and their evaluation does not fall within the objectives of this chapter; however, we point out the existence of two useful resources to perform both these operations, which can be quite heavy from the computational point of view: the Michigan Imputation Server (https://imputationserver.sph.umich.edu/index.html) [23] and the Sanger Imputation Service (https://imputation.sanger.ac.uk/) [24].

6. The number of PWMs can vary a lot between the available datasets and this can potentially influence the results of the analysis. However, in [11] we showed that the number of significant genes is not considerably affected by a substantial reduction in the number of PWMs used, possibly because of the redundancy inherent in larger PWMs databases, reflecting the similarities between the binding preferences of transcription factors.

7. In QTL mapping analysis covariates are commonly included in the fitted models in order to control for possible confounding factors. The set of covariates to be considered must be chosen based on the characteristics of each dataset; however, a fairly common goal is to correct for the population structure to better isolate the effects of individual variants from those of the genetic background. This goal is usually achieved by performing a principal component analysis of the genotypic data, and then including the top principal components (usually few of them are sufficient to explain the greatest amount of variation in the data) in the models as covariates. This is a quite simple step, but some precautions, such as LD-based pruning [13] and the exclusion of long-range LD regions [25], are recommended to avoid artifacts due to overweighting the regions of the genome with higher SNP density. Note that this filtering of genotypic data applies only to this step of the pipeline, while the whole dataset must be used for the computation of TBA values.

8. When vcf_rider is invoked using the -a argument it outputs a file that can be used to select the regions whose length changes due to indels. Every row has a region id, its start and end coordinates, and then a comma-separated list of all its

overlapping polymorphisms. Mutations are described by three fields separated by underscores: their start coordinate, their length (positive for deletions, negative for insertions and 1 for SNPs) and a boolean value, true for indels and false for SNPs.

9. The data structures used by vcf_rider to represent indels currently pose a limit on the number of indels that an individual can have on a regulatory region: it will incur in overflow errors if any individual has more than 64 indels with the minor allele on a single regulatory region. We are working on a future release without this limitation, but in the meantime we made available a tool (indel_stats, check on the vcf_rider repository for instructions on how to use it) to check if this happens in some of your data and remove the corresponding regulatory regions for which vcf_rider would report wrong scores.
10. In some cases, especially when using distal enhancers, the same regulatory sequence can be associated to more than one gene. In this case, one set of rows must be produced for each associated gene: these sets of rows will have the same TBA values, but different values of the gene expression and the gene-related part of the row identifier.
11. In linear models a set of explanatory variables is directly used to regress a dependent variable. For example, standard eQTL mapping analyses are based on the fitting of linear models in which the independent variables are a single genetic variant together with few covariates. However, here we have to deal with a high number of explanatory variables, i.e., the TBA values computed for all the TFs on a regulatory region, that would result in a serious overfitting issue. To overcome this problem, we decided to perform a principal component regression (PCR). In PCR the regressors are the top principal components of the explanatory variables; therefore, the dimensionality of the dataset can be arbitrarily reduced. Variable selection methods, such as lasso and ridge regression, are also commonly exploited to overcome the overfitting issue; however, in this case we prefer PCR because it is a strictly unsupervised procedure, that do not rely on the dependent variable to establish which features are included in the model, and because it allows a rigorous evaluation of the statistical significance of the model.
12. Given a significant TBA model, an obvious question is which are the most relevant TFs for the gene regulation. In [11] we proposed the fitting of further "univariate TBA models" as a possible way to get this type of information. However, these results must be interpreted with caution, considering, for example, the high degree of similarity between certain PWMs and the not always straightforward association between them and the TFs which are expressed and active in a given cell type.

References

1. Lander ES, Linton LM, Birren B et al (2001) Initial sequencing and analysis of the human genome. Nature 409:860–921. https://doi.org/10.1038/35057062
2. Venter JC, Adams MD, Myers EW et al (2001) The sequence of the human genome. Science 291:1304–1351. https://doi.org/10.1126/science.1058040
3. The 1000 Genomes Project Consortium, Gibbs RA, Boerwinkle E, et al. (2015) A global reference for human genetic variation. Nature 526:68–74. https://doi.org/10.1038/nature15393
4. Visscher PM, Wray NR, Zhang Q et al (2017) 10 years of GWAS discovery: biology, function, and translation. Am J Hum Genet 101:5–22. https://doi.org/10.1016/j.ajhg.2017.06.005
5. Gusev A, Lee SH, Trynka G et al (2014) Partitioning heritability of regulatory and cell-type-specific variants across 11 common diseases. Am J Hum Genet 95:535–552. https://doi.org/10.1016/j.ajhg.2014.10.004
6. Maurano MT, Humbert R, Rynes E et al (2012) Systematic localization of common disease-associated variation in regulatory DNA. Science 337:1190–1195. https://doi.org/10.1126/science.1222794
7. Hindorff LA, Sethupathy P, Junkins HA et al (2009) Potential etiologic and functional implications of genome-wide association loci for human diseases and traits. Proc Natl Acad Sci 106:9362–9367. https://doi.org/10.1073/pnas.0903103106
8. Cookson W, Liang L, Abecasis G et al (2009) Mapping complex disease traits with global gene expression. Nat Rev Genet 10:184–194. https://doi.org/10.1038/nrg2537
9. Stormo GD, Schneider TD, Gold L et al (1982) Use of the "Perceptron" algorithm to distinguish translational initiation sites in E. coli. Nucleic Acids Res 10:2997–3011
10. Foat BC, Morozov AV, Bussemaker HJ (2006) Statistical mechanical modeling of genome-wide transcription factor occupancy data by MatrixREDUCE. Bioinformatics 22:e141–e149. https://doi.org/10.1093/bioinformatics/btl223
11. Grassi E, Mariella E, Forneris M et al (2017) A functional strategy to characterize expression Quantitative Trait Loci. Hum Genet 136:1477–1487. https://doi.org/10.1007/s00439-017-1849-9
12. Price AL, Patterson NJ, Plenge RM et al (2006) Principal components analysis corrects for stratification in genome-wide association studies. Nat Genet 38:904–909. https://doi.org/10.1038/ng1847
13. Novembre J, Johnson T, Bryc K et al (2008) Genes mirror geography within Europe. Nature 456:98–101. https://doi.org/10.1038/nature07331
14. Danecek P, Auton A, Abecasis G et al (2011) The variant call format and VCFtools. Bioinformatics 27:2156–2158. https://doi.org/10.1093/bioinformatics/btr330
15. Chang CC, Chow CC, Tellier LC et al (2015) Second-generation PLINK: rising to the challenge of larger and richer datasets. Gigascience 4:7. https://doi.org/10.1186/s13742-015-0047-8
16. Khan A, Fornes O, Stigliani A et al (2018) JASPAR 2018: update of the open-access database of transcription factor binding profiles and its web framework. Nucleic Acids Res 46:D260–D266. https://doi.org/10.1093/nar/gkx1126
17. Kulakovskiy IV, Vorontsov IE, Yevshin IS et al (2018) HOCOMOCO: towards a complete collection of transcription factor binding models for human and mouse via large-scale ChIP-Seq analysis. Nucleic Acids Res 46:D252–D259. https://doi.org/10.1093/nar/gkx1106
18. Wickham H (2011) The split-apply-combine strategy for data analysis. J Stat Softw 40:1–29. https://doi.org/10.18637/jss.v040.i01
19. Corradin O, Saiakhova A, Akhtar-Zaidi B et al (2014) Combinatorial effects of multiple enhancer variants in linkage disequilibrium dictate levels of gene expression to confer susceptibility to common traits. Genome Res 24:1–13. https://doi.org/10.1101/gr.164079.113
20. Stegle O, Parts L, Piipari M et al (2012) Using probabilistic estimation of expression residuals (PEER) to obtain increased power and interpretability of gene expression analyses. Nat Protoc 7:500–507. https://doi.org/10.1038/nprot.2011.457
21. Browning SR, Browning BL (2011) Haplotype phasing: existing methods and new developments. Nat Rev Genet 12:703–714. https://doi.org/10.1038/nrg3054
22. Marchini J, Howie B (2010) Genotype imputation for genome-wide association studies. Nat Rev Genet 11:499–511. https://doi.org/10.1038/nrg2796
23. Das S, Forer L, Schönherr S et al (2016) Next-generation genotype imputation service and

methods. Nat Genet 48:1284–1287. https://doi.org/10.1038/ng.3656

24. McCarthy S, Das S, Kretzschmar W et al (2016) A reference panel of 64,976 haplotypes for genotype imputation. Nat Genet 48:1279–1283. https://doi.org/10.1038/ng.3643

25. Price AL, Weale ME, Patterson N et al (2008) Long-range LD can confound genome scans in admixed populations. Am J Hum Genet 83:132–135. https://doi.org/10.1016/J.AJHG.2008.06.005

Chapter 4

Identification and Quantification of Splicing Quantitative Trait Loci

Ankeeta Shah and Yang I. Li

Abstract

Most complex traits, including diseases, have a large genetic component. Identifying the genetic variants and genes underlying phenotypic variation remains one of the most important objectives of current biomedical research. Unlike Mendelian or familial diseases, which are usually caused by mutations in the coding regions of individual genes, complex diseases are thought to result from the cumulative effects of a large number of variants, of which, the vast majority are noncoding. Therefore, to discern the genetic underpinnings of a complex trait, we must first understand the impact of noncoding variation, which presumably affects gene regulation. In this chapter, we outline the recent progress made and methods used to discover putative regulatory regions associated with complex traits. We will specifically focus on mapping splicing quantitative trait loci (sQTL) using Yoruba samples from GEUVADIS as a motivating example.

Key words Genetic variation, Complex traits, Gene expression, Splicing, Quantitative trait loci (QTL), Splicing QTL (sQTL)

1 Introduction

Phenotypic variation between individuals can be qualitative or quantitative. Qualitative traits are discretely identifiable such that they can be analyzed by counts or ratios. In contrast, quantitative traits, such as height [1], or molecular phenotypes, such as gene expression levels, vary continuously and often follow a normal distribution. Most observable variation between individuals in a population tends to be quantitative. The field of quantitative genetics aims to understand the genetic factors that contribute to observed variation between individuals. The process of identifying genetic loci that are associated with phenotypic variation is known as quantitative trait loci (QTL) mapping [2] or QTL discovery. In the last 20 years, QTL mapping has been used extensively in a wide variety of biological contexts, including agriculture, medical genetics, evolution, and functional genomics.

Xinghua Mindy Shi (ed.), *eQTL Analysis: Methods and Protocols*, Methods in Molecular Biology, vol. 2082,
https://doi.org/10.1007/978-1-0716-0026-9_4, © Springer Science+Business Media, LLC, part of Springer Nature 2020

2 Materials

2.1 Software Prerequisites

1. STAR Version 2.6.
2. SAMtools Version 0.1.19.
3. Python Version 2.7.
4. R Version 3.3.3.
5. LeafCutter Version 0.2.7.
6. HTSlib, specifically, `FastQTL` requires `tabix` and `bzip`.
7. FastQTL Version 2.165.

2.2 Data Requirements

In order to begin splicing quantitative trait loci (sQTL) mapping, one needs two pieces of information:

1. **Genotype information** for the individuals that are being studied. Genotyping data is typically in variant calling file (VCF) format. In this example, we will use genotypes for Yoruba (YRI) individuals from the Genetic European Variation in Health and Disease (GEUVADIS) Consortium [3].
2. **Phenotype information**, in our case, splicing quantifications, for the individuals for which we have genotype information. We will outline ways to obtain splicing information from short read RNA-seq data, in our example consisting of 88 lymphoblastoid cell line (LCL) RNA-seq YRI samples from GEUVADIS. We will include more detailed information about how to process RNA-seq data using LeafCutter.

The example YRI data has already been mapped to the reference genome and junctions have been extracted. This data can be found on https://www.dropbox.com/s/ovn7cioi91bqpwi/sQTL_Chapter_junc.tar.gz?dl=0. We have subset on just chromosome 22 to make running through the tutorial fast and efficient.

3 Methods

Unlike linkage mapping [2], which is the approach that was traditionally used for QTL mapping of individual loci in which experimental crosses were performed or small pedigrees were used, association mapping of a sample of individuals from a population is now routine, particularly on a genome-wide scale (i.e. genome-wide association studies (GWAS)). This approach allows us to link common genetic variation associated with complex traits. The intuition behind this shift in approach is that, at the population level, a larger number of recombination events are able to reduce linkage disequilibrium (LD), which would normally restrict the resolution of our QTL mapping of individual loci. Therefore, association

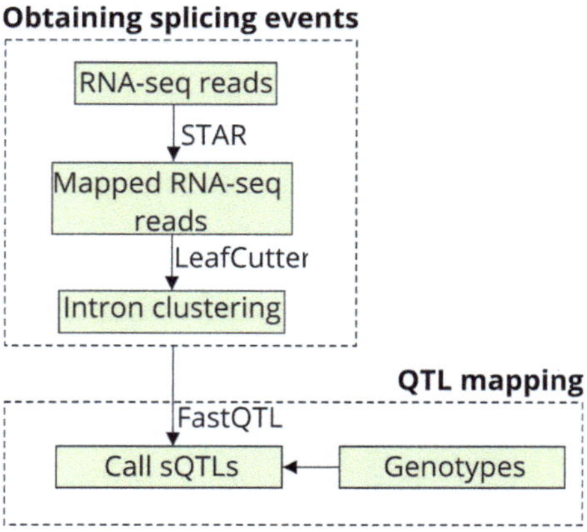

Fig. 1 Steps of sQTL analysis

studies allow for much finer resolution mapping of QTL. Nevertheless, identifying the precise molecular targets of these trait-associated genetic variants remains challenging.

In 2010, a seminal study by Nicolae and colleagues [4] showed that GWAS hits were highly enriched in expression quantitative trait loci (eQTLs), i.e., genetic variants that affect the expression level of a gene. This motivated a large number of studies that focused on mapping eQTLs in different tissues, cell-types, and contexts, in the hope of improving our understanding of how trait-associated variants mechanistically act to affect phenotypic traits. Recently, we found that RNA splicing is another major link between genetic variation and complex traits [5]. Briefly, RNA splicing is the process by which intronic sequences are removed from precursor messenger RNA (pre-mRNA). Mutations that disrupt splice sites within the pre-mRNA can result in the production of aberrant mRNA [6], which may ultimately lead to disease. In this chapter, we will highlight state of the art computational approaches that enable the rapid discovery and characterization of genetic variants that alter splicing (i.e., splicing quantitative trait loci, sQTLs (Fig. 1). However, many of these steps are applicable to other types of QTL analysis of molecular phenotypes, such as eQTL mapping.

3.1 Obtaining Splicing Events

RNA-seq reads represent fragments of mature mRNA species from which introns have been removed and exons have been ligated together. Therefore, it is possible to use this information in multiple ways to quantify RNA splicing. Some methods estimate mRNA isoform ratios [7–9] and others estimate exon inclusion levels [10, 11]. However, there are a number of statistical and technical issues that arise when using these types of methods to capture splicing events. In particular, the splicing events that can be captured by these methods may be limited because both of these quantification approaches rely on transcript models or annotated splicing events. Nevertheless, alternative methods have been developed, including MAJIQ [12], which can both identify novel splicing events and estimate inclusion levels without annotation.

Here, we describe `LeafCutter` [13], a method that we have recently developed, specifically for sQTL mapping. `LeafCutter` identifies and quantifies annotated and novel splicing events by focusing on intron excision events. The intuition behind this "intron-centric" method is that mRNA splicing occurs through the step-wise removal of introns from nascent pre-mRNA, resulting in ligated exon–exon junctions in the mature mRNA. `LeafCutter` uses the junction reads that are captured from RNA-seq, which are representative of intron splicing or excision events, to identify all possible junctions. Here, we will demonstrate how to use `LeafCutter` to obtain alternative splicing quantifications using the aforementioned example RNA-seq dataset.

3.1.1 Mapping RNA-seq Reads to a Reference Genome

Assuming that the standard quality control steps have been run on your RNA-seq samples, you can use `OLego` [14] or `STAR` [15] to align the RNA-seq dataset to a reference genome, which is hg19 in this example. `STAR` is advantageous to use because alignment is relatively fast as compared to other methods. `OLego` is also a nice alternative because it is particularly useful for finding de novo splice junctions.

Note that the example data we provide has already been mapped and junctions have been extracted. However, we include this step here for reference or in the case that you are using this protocol to analyze a different set of data. The raw LCL RNA-seq data for GEUVADIS can be found https://www.ebi.ac.uk/ena/data/view/PRJEB3366.

In this example, we will use `STAR`. `STAR` requires a genome index. To generate the index file using the GRCh38 or hg19 reference genome in FASTA format (`hg19.fa`) and annotations in GTF format `genocode_v19.gtf`:

```
STAR --runMode genomeGenerate --genomeDir hg19index/ --genomeFastaFiles
    hg19.fa --sjdbGTFfile gencode_v19.gtf --sjdbOverhang 100
```

1. `--runMode genomeGenerate` specifies that we are generating a genome index.
2. `--genomeDir hg19index/` is the directory in which the index will be output.
3. `--genomeFastaFiles hg19.fa` is the input hg19 reference genome in FASTA (or FASTQ) format.
4. `--sjdbGTFfile gencode_v19.gtf` contains the genome annotations in GTF format.
5. `--sjdbOverhang 100` specifies the length of the genomic sequence around the junction (in this case, 100 bp).

Then, RNA-seq reads (paired-end) can be mapped to the hg19 reference genome using:

```
STAR --genomeDir hg19index/ --twopassMode Basic --outSAMstrandField
    intronMotif --readFilesCommand RNAseqGeuvadis_1.fastq.gz
    RNAseqGeuvadis_2.fastq.gz --outSAMtype BAM Unsorted
```

1. `--genomeDir hg19index/` is the directory where the genome index is that was created in the previous step.
2. `--twopassMode Basic` specifies that STAR will run 2-pass mapping for each sample. STAR will perform the first pass to map the reads, extract junctions, and insert the junctions into the genome index. In the second pass, STAR will then re-map the reads.
3. `--outSAMstrandField intronMotif` is used when working with non-stranded data.
4. `--readFilesCommand RNAseqGeuvadis_1.fastq.gz RNAseqGeuvadis_2.fastq.gz` specifies are paired-end input RNA-seq reads, which should be in FASTA or FASTQ format.
5. `--outSAMtype BAM Unsorted` specifies our output alignment file format, which, in this case will be unsorted BAM files.

Considering Mapping Biases: Splicing quantification from short read RNA-seq data can be biased by polymorphisms that affect mapping of the reads. These are known as reads mapping with allelic bias (i.e. allele-specific reads), in which reads that carry the alternative allele generally have a lower probability of mapping correctly to the reference genome. Therefore, this read mapping bias can result in false positive QTLs identified as having associations with a complex trait [16]. These type of reads are important to remove when a variant is covered by reads that span intron junctions as this can lead to spurious association between the variant and counts associated with alternatively excised intron clusters. We recommend using WASP [17] to remove these.

3.1.2 Intron Clustering

As we have noted above, for the purpose of this tutorial, we are working with BAM files subset on chromosome 22. We did this by running:

```
for bamfile in 'ls *.bam'
do
samtools view -b $bamfile chr22 > $bamfile.chr22.bam
done
```

You can then convert these mapped reads to junctions, which are representative of introns that were removed from the nascent pre-RNA. These splicing events are our phenotypes that we want to associate with genotypes in subsequent steps.

```
for bamfile in 'ls *chr22.bam'
do
python bam2junc.py $bamfile $bamfile.junc
echo $bamfile.junc >> RNAseqGeuvadis_chr22_test_juncfiles.txt
done
```

One can then use `LeafCutter` to pool together all junctions with overlapping introns, which can be clustered together. They are demarcated by split reads. More specifically, overlapping introns that share a 3′ splice site or 5′ splice site are pooled together to form a cluster of introns, which represent alternative intron excision events. Noisy splicing events, otherwise described as rarely used introns with very little read overlap as compared to other introns in the same cluster, are filtered out. This can all be done by using the `RNAseqGeuvadis_chr22_test_juncfiles.txt` file, which is a list of all junction files that were created in the previous step, as input.

```
python leafcutter_cluster.py -j RNAseqGeuvadis_chr22_test_juncfiles.txt
    -m 50 -l 500000 -p MINCLURATIO -o RNAseqGeuvadis_chr22
```

1. `-j RNAseqGeuvadis_chr22_test_juncfiles.txt` contains a list of all junction files to be clustered.
2. `-m 50` specifies the minimum number of split reads in a cluster (default is 30 reads), but in this example, we specify 50 reads.
3. `-l 500000` specifies the maximum intron length in bp (default 100,000 bp), but in this example, we specify 500,000 bp (or 500 kb).
4. `-p MINCLURATIO` specifies the minimum fraction of reads in a cluster that support a junction (default 0.001).
5. `-o RNAseqGeuvadis_chr22` is the prefix of our output files (default "leafcutter").

With this, we now have a compilation of all introns that are alternatively excised. The output contains clusters of introns found in the junction files listed in `RNAseqGeuvadis_chr22_test_juncfiles.txt`, requiring, in this case, 50 split reads to support each cluster of up to 500 kb. The exact parameters can be modified depending on the particular coverage and window size you wish to work with. The output contains files that should be appended by `*_perind_numers.counts.gz`, in which every column corresponds to a different individual and every row corresponds to a different intron (identified by `chromosome:start:end:cluster_id`).

3.2 Discovering Genetic Variants Associated with Alternative Splicing

With our genotype information for all individuals and the clusters of alternatively excised introns we just generated, we can map sQTLs to understand how genetic variation affects alternative splicing. There are several methods for sQTL mapping, such as GLiMMPS [18], sQTLseekeR [19], and Altrans [20]. However, a major limitation of these methods is that they rely on existing isoform annotations. Therefore, for the purpose of this tutorial, we recommended using `LeafCutter` to obtain alternatively excised intron clusters, and, in the next step, `FastQTL` [21] to map sQTLs. In combination, these two pieces of software will allow us to identify more sQTLs overall than other existing methods because they are not restricted by using information about existing isoform annotations.

3.2.1 QTL Mapping

The goal of QTL discovery is to find statistically significant associations between genetic variants and phenotypes. This type of analysis, in the context of molecular QTL analysis, deals with thousands of molecular phenotypes that are simultaneously being measured. Therefore, one has millions of association tests that need to be run in order to scan all possible variant-phenotype combinations in *cis*. The standard approach is to perform single variant association testing for all variant-phenotype pairs, and for each locus, the minimum *p*-value from all member variants is then converted to the locus-level *p*-value. Unfortunately, the null distribution for the locus-level *p*-value is not known because of linkage disequilibrium. Therefore, permutation schemes have been developed to approximate the null distribution of locus-level *p*-values. `Matrix eQTL` [22], for example, utilizes a permutation scheme, and it has been used in multiple large-scale studies because it is able to perform matrix operations to efficiently run this large number of association tests with reasonable run-time. This permutation method can be computationally expensive, but `FastQTL`, as its name suggests, overcomes this computational cost with its own fast and efficient permutation scheme implementation, in which the null distribution of associations for a phenotype is modeled using a beta distribution. This makes it easy to approximate the tail of the distribution using only a handful of permutations such that it takes a short amount of time to estimate adjusted *p*-values at whatever significance level the user chooses. Subsequently, false discovery rate (FDR) control procedures can be applied to correct for multiple testing. Here, we will first outline how to find associations between genetic variants and phenotypes and then comment further on multiple testing correction.

1. **Prepare Phenotypes and Covariates** `LeafCutter` contains a script called `prepare_phenotype_table.py` that will normalize and standardize the intron usage ratios. Because linear regression is relatively sensitive to outliers, it is therefore important to remove the effects of outliers by performing normalization and standardization. The normalization works as outlined previously [23]. In brief, to meet the assumptions of the linear regression, the distribution of intron usage ratios are quantile-normalized within an individual to a standard normal distribution. Then, `prepare_phenotype_table.py` calculates N (default = 50) principal components (PCs) to be used as covariates in a linear regression. These PCs capture variance due to experimental differences between individuals and other confounding effects, and therefore, they are regressed out. Generally, these PCs do not capture *cis*-genotype effects because *cis*-genotype effects tend to be small and local. Thus, regressing out PCs will not remove our signal. In general, we suggest to use 10 phenotype PCs and 3 genotype PCs to remove confounders. Overall, the removal of these PCs, or covariates, will increase power to detect QTL, but, in general, will not decrease the false positive rates. However, the latter is generally not a problem as the number of individuals increases. These covariates are stored in files appended by `*_perind.counts.gz.PCs.gz`, and the files match `FastQTL` covariate input. In addition, this script also creates phenotype files separated by chromosome that can be used as input for `FastQTL`. In our case, because we are only working with chromosome 22, we will only have files that are appended as such `*_perind.counts.gz.qqnorm_chr22.gz`.

```
for file in 'ls *_perind.counts.gz'
do
python scripts/prepare_phenotype_table.py $file -p 10
done
```

`-p 10` specifies the number of PCs or covariates (default 50), but in this example, we specify 10.

`FastQTL` requires compressed and indexed phenotype files. `LeafCutter` generates a script (resulting from the above step) that one can run to do this: `sh *_perind.counts.gz_prepare.sh`.

Finally, one needs to compress and index the phenotype files:

```
for file in 'ls *_perind.counts.gz.qqnorm_chr22.gz'
done
bgzip -f $file
tabix -p bed $file
done
```

2. **Prepare Genotypes** One also needs to compress and index the genotype files, which can be found here:

```
bgzip -f
    GEUVADIS.chr22.PH1PH2_465.IMPFRQFILT_BIALLELIC_PH.annotv2.genotypes.vcf.gz
tabix -p vcf
    GEUVADIS.chr22.PH1PH2_465.IMPFRQFILT_BIALLELIC_PH.annotv2.genotypes.vcf.gz
```

3. **Mapping sQTLs** Next, one can run `FastQTL` to map sQTLs.

```
for file in 'ls *_perind.counts.gz.qqnorm_chr22.gz' | sed
    s'/_perind.counts.gz.qqnorm_chr22.gz//'g
do
FastQTL/bin/fastQTL.static --vcf
    GEUVADIS.chr22.PH1PH2_465.IMPFRQFILT_BIALLELIC_PH.annotv2.genotypes.vcf.gz
    --bed $file_perind.counts.gz.qqnorm_chr22.gz --permute 1000
    --chunk 1 1 --window 1e6 --cov $file_perind.counts.gz.PCs -out
    $file_output.txt.gz
```

(a) `--vcf GEUVADIS.chr22.PH1PH2_465.IMPFRQFILT_-BIALLELIC_PH.annotv2.genotypes.vcf.gz` contains genotype information for all individuals in VCF format.

(b) `--bed $file_perind.counts.gz.qqnorm_chr22.gz` contains phenotype (splicing) information for all individuals in BED format.

(c) `--permute 1000` specifies we want FastQTL to run permutations. We can omit this if we only want to identify nominal associations without running permutations.

(d) `-chunk 1 1` indicates we want to mapping on all of chromosome 22 at once. In theory, we could specify an alternate number of chunks to run jobs in parallel on a cluster, for example, `--chunk $j 30` where $j is replaced by

1 through 30 such that FastQTL will split up chromosome 22 into 30 regions so that we can run FastQTL separately on these 30 regions.

(e) `--window 1e6` specifies that we want to map QTL in a 1MB window.

(f) `--cov $file_perind.counts.gz.PCs` are our principal components (recall, we had 10 phenotype PCs and 3 genotype PCs to regress out from above).

(g) `-out $file_output` is our output file.

The output file will contain a nominal *p*-value, which signifies how significant from zero the regression coefficient is. Moreover, the output file will also contain permutation *p*-values, which are adjusted or corrected versions of the nominal *p*-value that take into account the fact that there are multiple genetic variants being tested per molecular phenotype.

3.2.2 Multiple Testing Correction

As we mentioned above, in order to draw valid conclusions from our QTL analysis, we need to associate statistical confidence measures with our observations. Subsequently, we must use a multiple testing correction procedure to adjust our statistical confidence measures based on the number of tests performed, in our case, the number of variant-phenotype pairs being tested in sQTL analysis. The *p*-values are corrected by the number of clusters we are testing using FDR control procedures, such as the Bonferroni correction [24], Benjamini-Hochberg [25], Storey's *q*-value [26].

This can be done by taking the output file from the mapping step and running the following command in R for Benjamini-Hochberg correction, for example:

```
d = read.table("$file_output.txt.gz", head=F, stringsAsFactors=F)
colnames(d) = c("pid", "nvar", "shape1", "shape2", "dummy", "sid",
    "dist", "npval", "slope", "ppval", "bpval")
d$bh = p.adjust(d$bpval, method="fdr")
write.table(d[which(d$bh <= 0.10), c(1,6)],
    "$f.permutations.all.chunks.benjamini.txt", quote=F, row.names=F,
    col.names=T)
```

We read in our output file from the previous step, add column names, correct for multiple testing using the Benjamini-Hochberg correction, and finally, write an output file that contains all sQTLs with a 10% false discovery rate (FDR). This output file will contain the first six columns of the original output step of FastQTL.

4 Conclusions

In this chapter, we have outlined steps you can take to discover and characterize genetic variants that alter splicing (i.e., sQTL analysis). Specifically, we have focused on mapping sQTLs in Yoruba samples from GEUVADIS as our motivating example of an application of this type of analysis. However, we emphasize that the steps presented in this tutorial are generalizable to other types of QTL analysis for molecular phenotypes (e.g., eQTL analysis). Ultimately, this type of analysis will allow us to better understand the genetic basis of complex traits, including disease.

References

1. Hirschhorn JN, Lindgren CM, Daly MJ, Kirby A, Schaffner SF, Burtt NP, Altshuler D, Parker A, Rioux JD, Platko J, *et al* (2001) Genomewide linkage analysis of stature in multiple populations reveals several regions with evidence of linkage to adult height. Am J Hum Genet 69:106–116
2. Kruglyak L, Lander ES (1995) A nonparametric approach for mapping quantitative trait loci. Genetics 139:1421–1428
3. Lappalainen T, Sammeth M, Friedländer MR, 't Hoen PAC, Monlong J, Rivas MA, Gonzalez-Porta M, Kurbatova N, Griebel T, Ferreira PG, *et al* (2013) Transcriptome and genome sequencing uncovers functional variation in humans. Nature 501:506
4. Nicolae DL, Gamazon E, Zhang W, Duan S, Dolan ME, Cox NJ (2010) Trait-associated SNPs are more likely to be eQTLs: annotation to enhance discovery from GWAS. PLoS Genet 6:e1000888
5. Li YI, van de Geijn B, Raj A, Knowles DA, Petti AA, Golan D, Gilad Y, Pritchard JK (2016) RNA splicing is a primary link between genetic variation and disease. Science 352:600–604
6. Pagani F, Baralle FE (2004) Genomic variants in exons and introns: identifying the splicing spoilers. Nat Rev Genet 5:389
7. Trapnell C, Hendrickson DG, Sauvageau M, Goff L, Rinn JL, Pachter L (2013) Differential analysis of gene regulation at transcript resolution with RNA-seq. Nat Biotechnol 31:46
8. Patro R, Mount SM, Kingsford C (2014) Sailfish enables alignment-free isoform quantification from RNA-seq reads using lightweight algorithms. Nat Biotechnol 32:462
9. Bray NL, Pimentel H, Melsted P, Pachter L (2016) Near-optimal probabilistic RNA-seq quantification. Nat Biotechnol 34:525
10. Katz Y, Wang ET, Airoldi EM, Burge CB (2010) Analysis and design of RNA sequencing experiments for identifying isoform regulation. Nat Methods 7:1009
11. Anders S, Reyes A, Huber W (2012) Detecting differential usage of exons from RNA-seq data. Genome Res 22:133744
12. Vaquero-Garcia J, Barrera A, Gazzara MR, Gonzalez-Vallinas J, Lahens NF, Hogenesch JB, Lynch KW, Barash Y (2016) A new view of transcriptome complexity and regulation through the lens of local splicing variations. Elife 5:e11752
13. Li YI, Knowles DA, Humphrey J, Barbeira AN, Dickinson SP, Im HK, Pritchard JK (2018) Annotation-free quantification of RNA splicing using LeafCutter. Nat Genet 50:151
14. Wu J, Anczukow O, Krainer AR, Zhang MQ, Zhang C (2013) OLego: fast and sensitive mapping of spliced mRNA-seq reads using small seeds. Nucleic Acids Res 41:5149–5163
15. Dobin A, Davis CA, Schlesinger F, Drenkow J, Zaleski C, Jha S, Batut P, Chaisson M, Gingeras TR (2013) Star: ultrafast universal RNA-seq aligner. Bioinformatics 29:15–21
16. Degner JF, Marioni JC, Pai AA, Pickrell JK, Nkadori E, Gilad Y, Pritchard JK (2009) Effect of read-mapping biases on detecting allele-specific expression from RNA-sequencing data. Bioinformatics 25:3207–3212
17. Van de Geijn B, McVicker G, Gilad Y, Pritchard JK (2015) WASP: allele-specific software for robust molecular quantitative trait locus discovery. Nat Methods 12:1061
18. Zhao K, Lu Zx, Park JW, Zhou Q, Xing Y (2013) GLiMMPS: robust statistical model for regulatory variation of alternative splicing using RNA-seq data. Genome Biol 14:R74

19. Monlong J, Calvo M, Ferreira PG, Guigó R (2014) Identification of genetic variants associated with alternative splicing using sQTLseekeR. Nat Commun 5:4698
20. Ongen H, Dermitzakis ET (2015) Alternative splicing QTLs in European and African populations. Am J Hum Genet 97:567–575
21. Ongen H, Buil A, Brown AA, Dermitzakis ET, Delaneau O (2015) Fast and efficient QTL mapper for thousands of molecular phenotypes. Bioinformatics 32:1479–1485
22. Shabalin AA (2012) Matrix eQTL: ultra fast eQTL analysis via large matrix operations. Bioinformatics 28:1353–1358
23. Degner JF, Pai AA, Pique-Regi R, Veyrieras JB, Gaffney DJ, Pickrell JK, De Leon S, Michelini K, Lewellen N, Crawford GE, *et al* (2012) DNase I sensitivity QTLs are a major determinant of human expression variation. Nature 482:390
24. Bonferroni, C (1936) Teoria statistica delle classi e calcolo delle probabilita. Pubblicazioni del R Istituto Superiore di Scienze Economiche e Commericiali di Firenze 8:3–62
25. Benjamini Y, Hochberg Y (1995) Controlling the false discovery rate: a practical and powerful approach to multiple testing. J R Stat Soc Ser B 57:289–300
26. Storey JD, Tibshirani R (2003) Statistical significance for genomewide studies. Proc Natl Acad Sci 100:9440–9445

Chapter 5

Genome-Wide Composite Interval Mapping (GCIM) of Expressional Quantitative Trait Loci in Backcross Population

Yuan-Ming Zhang

Abstract

One of the most remarkable findings in expressional quantitative trait locus (eQTL) mapping is that *trans* (distal) eQTL has small effect. The widely used approaches have a low power in the detection of small-effect eQTL. To overcome this issue, we integrate polygenic background control with multi-locus genetic model to develop genome-wide composite interval mapping (GCIM). This chapter covers the GCIM procedure in a backcross or doubled haploid populations. We describe the genetic model, parameter estimation, multi-locus genetic model, hypothesis tests, and software. Finally, some issues related to the GCIM method are discussed.

Key words Expressional quantitative trait locus, Backcross, Genome-wide composite interval mapping, Empirical Bayes, Multi-QTL model, Wald test

1 Introduction

Genetical genomics is the integration of genomics with genetics [1]. In the genetical genomics, expressional levels are viewed as quantitative traits and previously quantitative trait locus (QTL) mapping approaches are used to detect their genetic variants that affect gene regulation. This is expressional QTL (eQTL) mapping. The most remarkable findings in eQTL studies are that regulatory variation in gene expression level is highly heritable and most strong eQTL are found to be near the target gene [2]. The eQTL consist of *cis* (proximal) or *trans* (distal) eQTL, which are based on their physical distances from the irregulated genes (proximal regions are defined as a 2-MB window containing the gene), although the distance-based classification of eQTL as *cis* or *trans* requires empirical validation. In these two kinds of eQTL, up to one-third of eQTL are *cis* acting [3]. In most eQTL mapping studies, relatively few lines are analyzed due to the expense of

Xinghua Mindy Shi (ed.), *eQTL Analysis: Methods and Protocols*, Methods in Molecular Biology, vol. 2082,
https://doi.org/10.1007/978-1-0716-0026-9_5,

genome-wide measurements. This severely limits the ability to detect eQTL with small effects [2], especially for *trans* eQTL. To increase the power for eQTL detection, statistical methods need to be addressed.

Currently, the widely used approach in QTL mapping methodologies is composite interval mapping (CIM) [4–6], although inclusive CIM [7] and Bayesian methodologies [8, 9] can improve the power and accuracy of QTL detection. However, all the above methods have a low power in the detection of small and linked QTL, especially for the linked QTL with effects in opposite directions. Recently, we have integrated the polygenic background control of genome-wide association studies with the multi-locus genetic model method to establish a new framework of multi-locus QTL mapping [10]. This approach calls genome-wise CIM (GCIM), which can detect small and linked QTL. Most *trans* eQTL have small effects, and GCIM may be used to detect these *trans* eQTL. In this chapter, we focus on the GCIM method [10].

2 Genome-Wide Composite Interval Mapping

2.1 Genetic Model

Let $\mathbf{y}$ be a $n \times 1$ vector of phenotypic values for n individuals in a backcross or doubled haploid (DH) populations. Let $\mathbf{Z}_k$ be a genotype indictor variable for marker k, where the jth element of $\mathbf{Z}_k$ is defined as

$$Z_{jk} = \begin{cases} +1 \text{ for } AA \\ -1 \text{ for } Aa \end{cases} \quad (1)$$

We now write the model by

$$\mathbf{y} = \mathbf{X}\boldsymbol{\alpha} + \mathbf{Z}_k\gamma_k + \boldsymbol{\xi} + \boldsymbol{\varepsilon} \quad (2)$$

where $\mathbf{X}$ is a design matrix for (non-genetic) fixed effects $\boldsymbol{\alpha}$; $\gamma_k \sim N(0, \phi_k^2)$ is the effect of marker k and ϕ_k^2 is estimated from the data; $\boldsymbol{\xi}$~MVN($\mathbf{0}$, $\mathbf{K}\phi^2$) is a $n \times 1$ polygenic effect vector, kinship matrix $\mathbf{K}$ is calculated using genome-wide marker information, and MVN denotes multivariate normal distribution; and the random effect $\boldsymbol{\varepsilon}$~MVN($\mathbf{0}$, $\mathbf{I}\sigma^2$) is the residual error, and $\mathbf{I}$ is a $n \times n$ identity matrix.

If marker density is less than 1 cM, it is enough to scan marker positions on the genome. If not, we insert one pseudo marker in every d cM to cover the entire genome evenly so that every position of the genome will be evaluated. When a pseudo marker is located between two consecutive markers, multipoint method [11] may be used to calculate the genotype probabilities, denoted by $p_{jk}(AA)$ and $p_{jk}(Aa)$, respectively, for the two genotypes AA and Aa in backcross. The genotype indicator variable, Z_{jk}, is then defined as the expected value conditional on flanking marker genotypes [12]. Therefore, Z_{jk} is defined as

$$Z_{jk} = (+1)p_{jk}(AA) + (-1)p_{jk}(Aa) = p_{jk}(AA) - p_{jk}(Aa) \quad (3)$$

The expectation of $\mathbf{y}$ is $\mathrm{E}(\mathbf{y}) = \mathbf{X\alpha}$ and the variance is

$$\begin{aligned} var(\mathbf{y}) &= \mathbf{Z}_k\mathbf{Z}_k^T\phi_k^2 + \mathbf{K}\phi^2 + \mathbf{I}\sigma^2 \\ &= \left[\mathbf{Z}_k\mathbf{Z}_k^T\lambda_k + (\mathbf{K}\lambda + \mathbf{I})\right]\sigma^2 \\ &= \left(\mathbf{Z}_k\mathbf{Z}_k^T\lambda_k + \mathbf{H}\right)\sigma^2 \end{aligned} \quad (4)$$

where $\lambda_k = \phi_k^2/\sigma^2$, $\lambda = \phi^2/\sigma^2$, $\mathbf{H} = \mathbf{K}\lambda + \mathbf{I}$, and

$$\mathbf{K} = \frac{1}{m}\sum_{k=1}^{m}\mathbf{Z}_k\mathbf{Z}_k^T \quad (5)$$

is a marker-inferred kinship matrix [13].

2.2 Parameter Estimation

We first considered the reduced form of the model (2), which deleted $\mathbf{Z}_k\gamma_k$,

$$\mathbf{y} = \mathbf{X\alpha} + \boldsymbol{\xi} + \boldsymbol{\varepsilon} \quad (6)$$

The variance of $\mathbf{y}$ is:

$$var(\mathbf{y}) = \phi^2\mathbf{K} + \sigma^2\mathbf{I} = \sigma^2(\lambda\mathbf{K} + \mathbf{I}) \quad (7)$$

Using efficient mixed model association (EMMA) algorithm [14], the estimate of λ, denoted by $\hat{\lambda}$, can be easily obtained.

We then considered the model (2), and replaced λ in Eq. 4 by $\hat{\lambda}$, so

$$var(\mathbf{y}) = \left[\mathbf{Z}_k\mathbf{Z}_k^T\lambda_k + \left(\mathbf{K}\hat{\lambda} + \mathbf{I}\right)\right]\sigma^2 \quad (8)$$

An Eigen (or spectral) decomposition of matrix $\mathbf{K}$ is $\mathbf{U}^T\mathbf{K}\mathbf{U} = \mathbf{D}$. Let $\mathbf{y}^* = \mathbf{U}^T\mathbf{y}$, $\mathbf{X}^* = \mathbf{U}^T\mathbf{X}$, and $\mathbf{Z}_k^* = \mathbf{U}^T\mathbf{Z}_k$ be transformed variables so that

$$\mathbf{y}^* = \mathbf{X}^*\boldsymbol{\alpha} + \mathbf{Z}_k^*\gamma_k + \mathbf{U}^T(\boldsymbol{\xi} + \boldsymbol{\varepsilon}) \quad (9)$$

The variance-covariance matrix of $\mathbf{y}^*$ is

$$\begin{aligned} var(\mathbf{y}^*) &= \mathbf{Z}_k^*\mathbf{Z}_k^{*\mathsf{T}}\phi_k^2 + \mathbf{U}^T var(\boldsymbol{\xi})\mathbf{U} + \mathbf{I}\sigma^2 \\ &= \mathbf{Z}_k^*\mathbf{Z}_k^{*\mathsf{T}}\phi_k^2 + \mathbf{U}^T\mathbf{K}\mathbf{U}\phi^2 + \mathbf{I}\sigma^2 \\ &= \mathbf{Z}_k^*\mathbf{Z}_k^{*\mathsf{T}}\phi_k^2 + \mathbf{D}\phi^2 + \mathbf{I}\sigma^2 \\ &= \mathbf{Z}_k^*\mathbf{Z}_k^{*\mathsf{T}}\phi_k^2 + \left(\mathbf{D}\hat{\lambda} + \mathbf{I}\right)\sigma^2 \\ &= \left(\mathbf{Z}_k^*\mathbf{Z}_k^{*\mathsf{T}}\lambda_k + \mathbf{R}\right)\sigma^2 \end{aligned} \quad (10)$$

where $\mathbf{R} = \mathbf{D}\hat{\lambda} + \mathbf{I}$ is a known diagonal matrix. Let $\mathbf{R}_k = \mathbf{Z}_k^*\mathbf{Z}_k^{*\mathsf{T}}\lambda_k + \mathbf{R}$ be the general covariance structure. After absorbing

$\boldsymbol{\alpha}$ and σ^2, we have the following profiled restricted log likelihood function,

$$L(\lambda_k) = -\frac{1}{2}\ln|\mathbf{R}_k| - \frac{1}{2}\ln|\mathbf{X}^{*\mathrm{T}}\mathbf{R}_k^{-1}\mathbf{X}^*| - \frac{n-r}{2}\ln\left(\mathbf{y}^{*\mathrm{T}}\mathbf{P}_k\mathbf{y}^*\right) \quad (11)$$

where

$$\mathbf{P}_k = \mathbf{R}_k^{-1} - \mathbf{R}_k^{-1}\mathbf{X}^*\left(\mathbf{X}^{*\mathrm{T}}\mathbf{R}_k^{-1}\mathbf{X}^*\right)^{-1}\mathbf{X}^{*\mathrm{T}}\mathbf{R}_k^{-1} \quad (12)$$

$r =$ rank $(\mathbf{X}^*)$. This likelihood function contains only one unknown parameter, λ_k. The Newton algorithm for λ_k is

$$\lambda_k^{(t+1)} = \lambda_k^{(t)} - \left[\frac{\partial^2 L\left(\lambda_k^{(t)}\right)}{\partial\lambda_k^2}\right]^{-1}\left[\frac{\partial L\left(\lambda_k^{(t)}\right)}{\partial\lambda_k}\right] \quad (13)$$

Once the iteration process converges, the solution is the REML estimate of λ_k, denoted by $\hat{\lambda}_k$. Given $\lambda_k = \hat{\lambda}_k$, the estimates of $\boldsymbol{\alpha}$ and σ^2 are

$$\begin{aligned}\hat{\boldsymbol{\alpha}} &= \left(\mathbf{X}^{*\mathrm{T}}\mathbf{R}_k^{-1}\mathbf{X}^*\right)^{-1}\mathbf{X}^{*\mathrm{T}}\mathbf{R}_k^{-1}\mathbf{y}^* \\ \hat{\sigma}^2 &= \frac{1}{n-r}(\mathbf{y}^* - \mathbf{X}^*\hat{\boldsymbol{\alpha}})^T\mathbf{R}_k^{-1}(\mathbf{y}^* - \mathbf{X}^*\hat{\boldsymbol{\alpha}})\end{aligned} \quad (14)$$

The best linear unbiased prediction (BLUP) of γ_k is also the conditional expectation of γ_k given $\mathbf{y}^*$ and has the following expression,

$$\begin{aligned}\mathrm{E}(\gamma_k|\mathbf{y}^*) &= \lambda_k\mathbf{Z}_k^{*\mathrm{T}}\mathbf{R}_k{}^{-1}\mathbf{y}^* \\ &\quad - \lambda_k\mathbf{Z}_k^{*\mathrm{T}}\mathbf{R}_k{}^{-1}\mathbf{X}^*\left(\mathbf{X}^{*\mathrm{T}}\mathbf{R}_k{}^{-1}\mathbf{X}^*\right)^{-1}\mathbf{X}^{*\mathrm{T}}\mathbf{R}_k{}^{-1}\mathbf{y}^* \quad (15)\end{aligned}$$

The conditional variance is

$$var(\gamma_k|\mathbf{y}^*) = \lambda_k\sigma^2 - \lambda_k\mathbf{Z}_k^{*\mathrm{T}}\mathbf{R}_k{}^{-1}\mathbf{Z}_k^*\lambda_k\sigma^2 \quad (16)$$

Under the random model approach, we first estimate the polygenic variance. We then estimate λ_k and test $\lambda_k = 0$ for each marker using the same estimated polygenic variance ratio.

2.3 Multi-Locus Random-QTL-Effect Mixed Linear Model Method

The method described in Subheadings 2.1 to 2.3 calls single-marker random-QTL-effect mixed linear model (rMLM) method. This approach is considered as an initial scanning step for multi-locus rMLM (mrMLM) method. In the rMLM step, the negative logarithm *P*value curve is obtained for each true or pseudo marker. In the curve, all the peaks were selected as the positions of putative QTL in a multi-locus QTL mapping model. Thus, multi-locus genetic model under consideration is as follows

$$\mathbf{y} = \mathbf{X}\boldsymbol{\alpha} + \sum_{k=1}^{K} \mathbf{Z}_k \gamma_k + \boldsymbol{\varepsilon} \tag{17}$$

where K is the number of peaks in the negative logarithm P value curve, $\boldsymbol{\gamma} = (\gamma_1, \gamma_2, \ldots, \gamma_K)^T$, and the others are same as those in the model (2). Note that $P(\gamma_k) \propto N(0, \sigma_k^2)$ and other prior distributions are same as those in [15].

All the effects of QTL in the multi-locus model were estimated by empirical Bayes [15]. The procedure for parameter estimation is as follows.

1. Setting initial values:

$$\begin{aligned} \sigma_1^2 &= \cdots = \sigma_K^2 = 1 \\ \boldsymbol{\alpha} &= \left(\mathbf{X}^T\mathbf{X}\right)^{-1}\mathbf{X}^T\mathbf{y} \\ \sigma^2 &= \frac{1}{2n}(\mathbf{y} - \mathbf{X}\boldsymbol{\alpha})^T(\mathbf{y} - \mathbf{X}\boldsymbol{\alpha}) \end{aligned} \tag{18}$$

2. E-step: QTL effect can be predicted by

$$\mathrm{E}(\gamma_k) = \sigma_k^2 \mathbf{Z}_k^T \mathbf{V}^{-1}(\mathbf{y} - \mathbf{X}\boldsymbol{\alpha}) \tag{19}$$

where $\mathbf{V} = \sum_{k=1}^{K} \mathbf{Z}_k \mathbf{Z}_k^T \sigma_k^2 + \mathbf{I}\sigma^2$;

3. M-step: update parameters σ_k^2, $\boldsymbol{\alpha}$, and σ^2

$$\begin{aligned} \sigma_k^2 &= \frac{\mathrm{E}\left(\gamma_k^T \gamma_k\right) + \omega}{\tau + 2 + 1} \\ \boldsymbol{\alpha} &= \left(\mathbf{X}^T\mathbf{V}^{-1}\mathbf{X}\right)^{-1}\mathbf{X}^T\mathbf{V}^{-1}\mathbf{y} \\ \sigma^2 &= \frac{1}{n}(\mathbf{y} - \mathbf{X}\boldsymbol{\alpha})\left(\mathbf{y} - \mathbf{X}\boldsymbol{\alpha} - \sum_{k=1}^{K} \mathbf{Z}_k \mathrm{E}(\gamma_k)\right) \end{aligned} \tag{20}$$

where $\mathrm{E}\left(\gamma_k^T \gamma_k\right) = \mathrm{E}\left(\gamma_k^T\right)\mathrm{E}(\gamma_k) + \mathrm{tr}[var(\gamma_k)]$, $var(\gamma_k) = \mathbf{I}\sigma_k^2 - \sigma_k^2 \mathbf{Z}_k^T \mathbf{V}^{-1} \mathbf{Z}_k \sigma_k^2$, and $(\tau, \omega) = (0, 0)$. Repeat E-step and M-step until convergence is satisfied.

2.4 Wald Test for Marker Effect

We use the Wald test to test $H_0 : \gamma_k = 0$ in model (17) and the Wald test statistic is

$$W_k = \frac{\hat{\gamma}_k^2}{var(\gamma_k)} \tag{21}$$

where $\hat{\gamma}_k$ and $\mathrm{var}(\gamma_k)$ are obtained from Eqs. 15 and 16, respectively [15].

2.5 Likelihood Ratio Test (LRT)

If QTL effect estimate in model (17) is apart from zero and its P value in the above Wald test is less than 0.01, the putative QTL are picked up to conduct LRT. All the putative QTL with the $\geq$2.5

LOD score were viewed as true. Considering that all potential QTL were selected in the first stage, we decided to place a slightly more stringent criterion of 0.000691, which is converted from LOD score 2.50 of the test statistics using $p = Pr(\chi^2_{v=1} > 2.50 \times 4.605) = 0.000691$.

2.6 R Software

Expressional levels for each gene are viewed as a quantitative trait and GCIM is used to detect its genetic variants that affect gene regulation. GCIM can be conducted by the programs "QTL.gCIMapping" or "QTL.gCIMapping.GUI," which are downloaded, respectively, from https://cran.r-project.org/web/packages/QTL.gCIMapping/index.html or https://cran.r-project.org/web/packages/QTL.gCIMapping.GUI/index.html.

3 Notes

1. The current method is suitable for backcross, doubled haploid, and recombinant inbred line population because there are only two genotypes in these populations. This situation is different from that in F_2 population, which has three genotypes, *AA*, *Aa*, and *aa*. This lets us estimate two kinds of effects (additive and dominant effects). Therefore, five kinds of variance components (additive, dominant, additive polygenes, dominant polygenes, and residual error) need to be estimated. This increases the difficulty in the estimation of variance components and running time. To overcome this issue, additive and dominant effects for each putative QTL are separately scanned so that a negative logarithm *P* value curve against genome position can be separately obtained for each effect. In each curve, all the peaks are viewed as potential QTL. Clearly, the major contribution of the GCIM algorithm in F_2 is to propose a statistical framework jointly using CIM, random model, and lasso techniques [16], and its R software has been incorporated into the programs "QTL.gCIMapping.GUI" and "QTL.gCIMapping."

   We need to explain why the new method is called GCIM. First, in GCIM we scan each position on the genome under polygenic background control, and this idea is similar to that in CIM. Second, "genome-wide" is derived from two meanings. One is to scan the whole genome and the other is to control all the background effects in the genome, including large, middle, and minor background effects. In CIM, it is impossible to control minor and some middle background effects. This may be the reason why the new method has higher power and accuracy in QTL detection than CIM.

2. Multi-locus QTL mapping has become the state-of-the-art procedure to identify QTL associated with quantitative traits. Thus, multiple interval mapping [17] has been proposed. In the case of a large number of QTLs, however, the running time is relatively long. Although Bayesian methodology is available in multiple QTL mapping, the same problem also appears. Moreover, it is hard to detect small and linked QTL, and a large number of markers make cofactor selection infeasible. To overcome these shortcomings, a random-QTL-effect mixed linear model framework of GWAS has been developed to identify QTL in backcross [13], but the method is not always significantly better than CIM in our Monte Carlo simulation experiments [10]. In our GCIM, all the polygenic effects are fitted to a mixed linear model to capture the genomic background information and all the peaks in the genome-wide scan curve are selected as the positions of multiple putative QTL so that GCIM has significantly improved the statistical power and accuracy of QTL detection.
3. In human genetics, genome-wide association studies would identify small genetic effects more efficiently than linkage analysis [18]. Since the mixed linear model (MLM) method by fitting a population structure (Q) and polygenic background (K) has been established [19, 20], many MLM-based methods have been proposed [14, 21–24]. However, these single-locus genome scan approaches need multiple test correction. The standard Bonferroni correction is often too conservative so that many significant loci usually do not pass the stringent criterion of significance test. To solve this issue, multi-locus GWAS methodologies have been developed [25, 26]. Actually, the two approaches are simple, stepwise mixed-model regression with forward inclusion and backward elimination, and the computationally intensive forward-backward inclusion of SNPs is clearly a limiting factor in exploring the huge model space [25]. To solve this issue, we have established a series of two-stage multi-locus GWAS methodologies [27–31]. In the first stage, all the markers on the genome are scanned by single-locus genome scan approach and all the potentially associated markers are selected. In the second stage, all the selected markers are placed into one genetic model, their effects are estimated by empirical Bayes, and all the nonzero effects are further detected by likelihood ratio test. The results from our methods are better than those from GEMMA, indicating the effectiveness of this strategy.

Acknowledgments

This work was supported by the National Natural Science Foundation of China (31571268, U1602261, 31871242), and Huazhong Agricultural University Scientific & Technological Self-innovation Foundation (Program No. 2014RC020).

References

1. Jansen RC, Nap JP (2001) Genetical genomics: the added value from segregation. Trends Genet 17(7):388–391
2. Gilad Y, Rifkin SA, Pritchard JK (2008) Revealing the architecture of gene regulation: the promise of eQTL studies. Trends Genet 24 (8):408–415
3. Gibson G, Weir B (2005) The quantitative genetics of Transcription. Trends Genet 21 (11):616–623
4. Zeng ZB (1993) Theoretical basis for separation of multiple linked gene effects in mapping of quantitative trait loci. Proc Natl Acad Sci U S A 90:10972–10976
5. Zeng ZB (1994) Precision mapping of quantitative trait loci. Genetics 136:1457–1468
6. Jansen RC (1993) Interval mapping of multiple quantitative trait loci. Genetics 135:205–211
7. Li HH, Ye GY, Wang JK (2007) A modified algorithm for the improvement of composite interval mapping. Genetics 175:361–374
8. Xu S (2003) Estimating polygenic effects using markers of the entire genome. Genetics 163:789–801
9. Wang H, Zhang YM, Li X, Masinde GL, Mohan S, Baylink DJ, Xu S (2005) Bayesian shrinkage estimation of QTL parameters. Genetics 170:465–480
10. Wang SB, Wen YJ, Ren WL, Ni YL, Zhang J, Feng JY, Zhang YM (2016) Mapping small-effect and linked quantitative trait loci for complex traits in backcross or DH populations via a multi-locus GWAS methodology. Sci Rep 6:29951
11. Jiang C, Zeng ZB (1997) Mapping quantitative trait loci with dominant and missing markers in various crosses from two inbred lines. Genetica 101:47–58
12. Haley CS, Knott SA (1992) A simple regression method for mapping quantitative trait loci in the line crosses using flanking markers. Heredity 69:315–324
13. Xu S (2013) Mapping quantitative trait loci by controlling polygenic background effects. Genetics 195:1209–1222
14. Kang HM, Zaitlen NA, Wade CM, Kirby A, Heckerman D, Daly MJ, Eskin E (2008) Efficient control of population structure in model organism association mapping. Genetics 178:1709–1723
15. Xu S (2010) An expectation–maximization algorithm for the Lasso estimation of quantitative trait locus effects. Heredity 105:483–494
16. Wen YJ, Zhang YW, Zhang J, Feng JY, Dunwell JM, Zhang YM (2018) An efficient multi-locus mixed model framework for the detection of small and linked QTLs in F_2. Brief Bioinform. https://doi.org/10.1093/bib/bby058
17. Kao CH, Zeng ZB, Teasdale RD (1999) Multiple interval mapping for quantitative trait loci. Genetics 152:1203–1216
18. Risch N, Merikangas K (1996) The future of genetic studies of complex human diseases. Science 273:1516–1517
19. Zhang YM, Mao Y, Xie C, Smith H, Luo L, Xu S (2005) Mapping quantitative trait loci using naturally occurring genetic variance among commercial inbred lines of maize (*Zea mays* L.). Genetics 169:2267–2275
20. Yu J, Pressoir G, Briggs WH, Vroh Bi I, Yamasaki M, Doebley JF, McMullen MD, Gaut BS, Nielsen DM, Holland JB, Kresovich S, Buckler ES (2006) A unified mixed-model method for association mapping that accounts for multiple levels of relatedness. Nat Genet 38:203–208
21. Zhang Z, Ersoz E, Lai CQ, Todhunter RJ, Tiwari HK, Gore MA, Bradbury PJ, Yu J, Arnett DK, Ordovas JM, Buckler ES (2010) Mixed linear model approach adapted for genome-wide association studies. Nat Genet 42:355–360
22. Zhou X, Stephens M (2012) Genome-wide efficient mixed-model analysis for association studies. Nat Genet 44:821–824
23. Zhou X, Carbonetto P, Stephens M (2013) Polygenic modeling with Bayesian sparse linear mixed models. PLoS Genet 9:e1003264
24. Li M, Liu X, Bradbury P, Yu J, Zhang YM, Todhunter RJ, Buckler ES, Zhang Z (2014)

Enrichment of statistical power for genome-wide association studies. BMC Biol 12:73

25. Segura V, Vilhjálmsson BJ, Platt A, Korte A, Seren Ü, Long Q, Nordborg M (2012) An efficient multi-locus mixed-model approach for genome-wide association studies in structured populations. Nat Genet 44:825–830
26. Liu X, Huang M, Fan B, Buckler ES, Zhang Z (2016) Iterative usage of fixed and random effect models for powerful and efficient genome-wide association studies. PLoS Genet 12(2):e1005767
27. Wang SB, Feng JY, Ren WL, Huang B, Zhou L, Wen YJ, Zhang J, Dunwell JM, Xu S, Zhang YM (2016) Improving power and accuracy of genome-wide association studies via a multi-locus mixed linear model methodology. Sci Rep 6:19444
28. Wen YJ, Zhang H, Ni YL, Huang B, Zhang J, Feng JY, Wang SB, Dunwell JM, Zhang YM, Wu R (2018) Methodological implementation of mixed linear models in multi-locus genome-wide association studies. Brief Bioinform 19 (4):700–712
29. Tamba CL, Ni YL, Zhang YM (2017) Iterative sure independence screening EM-Bayesian LASSO algorithm for multi-locus genome-wide association studies. PLoS Comput Biol 13(1):e1005357
30. Zhang J, Feng JY, Ni YL, Wen YJ, Niu Y, Tamba CL, Yue C, Song Q, Zhang YM (2017) pLARmEB: Integration of least angle regression with empirical Bayes for multi-locus genome-wide association studies. Heredity 118:517–524
31. Ren WL, Wen YJ, Dunwell JM, Zhang YM (2018) pKWmEB: integration of Kruskal–Wallis test with empirical Bayes under polygenic background control for multi-locus genome-wide association study. Heredity 120:208–218

Chapter 6

Combining eQTL and SNP Annotation Data to Identify Functional Noncoding SNPs in GWAS Trait-Associated Regions

Stephen A. Ramsey, Zheng Liu, Yao Yao, and Benjamin Weeder

Abstract

We describe a statistical method for prioritizing candidate causal noncoding single nucleotide polymorphisms (SNPs) in regions of the genome that are detected as trait-associated in a population-based genome-wide association study (GWAS). Our method's key step is to combine, within a naïve Bayes-like framework, three quantities for each SNP: (1) the *p*-value for the association test between the SNP's genotype and the trait; (2) the *p*-value for the SNP's *cis*-expression quantitative trait locus (*cis*-eQTL) association test; and (3) a model-based prediction score for the SNP's potential to be a regulatory SNP (rSNP). The method is flexible with respect to the source of the model-based rSNP prediction score; we demonstrate the method using scores obtained using the previously published machine-learning-based rSNP prediction method, CERENKOV2. Because it requires only the GWAS trait association test *p*-value for each SNP and not full genotype information, our method is applicable for GWAS secondary analysis in the common situation where only summary data (and not full genotype data) are readily available. We illustrate how the method works in step-by-step fashion.

Key words SNP, Noncoding, GWAS, Computational, eQTL, rSNP, Likelihood, Bioinformatics

1 Introduction

1.1 Background

Human genome-wide association studies (GWAS) have led to the identification of many genetic loci that are causal for population variation in various traits [32]; however, each of these loci typically has many trait-associated SNPs that are in strong linkage disequilibrium with one another across the population, making it difficult to the pinpoint causal SNP. Ranking candidate causal SNPs that are in coding regions is relatively straightforward because such SNPs' alleles can be mapped to consequence predictions based on the SNPs' predicted effects on the polypeptide chain [29]. However, 90% of human GWAS trait-associated SNPs are located in *noncoding* regions [17]. Within a noncoding trait-associated region, it is at present difficult to pinpoint the regulatory SNP (or rSNP) that is

Xinghua Mindy Shi (ed.), *eQTL Analysis: Methods and Protocols*, Methods in Molecular Biology, vol. 2082,
https://doi.org/10.1007/978-1-0716-0026-9_6,

causal for trait variation [30]. Thus, statistical and computational approaches have been proposed to integrate various types of information about noncoding SNPs in order to prioritize—within a trait-associated locus—the noncoding SNPs that are most likely be causal for the trait variation. Approaches to this problem fall into two categories: statistical fine-mapping strategies that co-analyze SNP genotypes within a locus (*see* [28] for a recent review), and computational approaches that fuse various types of SNP data (other than genotype) to rank noncoding SNPs for possible regulatory function. The "data fusion" approaches can be divided into "unsupervised" approaches [1, 6–8, 11, 13, 16, 25, 27, 33, 34] (which are not trained using a set of ground-truth experimentally validated noncoding SNPs) and approaches that are "supervised" (i.e., trained) using data from an example set of experimentally validated rSNPs or GWAS noncoding SNPs [4, 14, 19, 23, 24, 26]. The types of beneficial SNP data include population genetic information (e.g., allele frequency), phylogenetic information (e.g., interspecies sequence conservation scores), nearest-gene functional annotations, locus replication timing data, three-dimensional chromatin interaction data, and expression quantitative trait locus (eQTL) data. Converging lines of evidence [4, 14, 20] have underscored that eQTL information is an important resource for data fusion methods to identify causal noncoding SNPs, due to the fact that such methods link noncoding SNPs to gene expression changes within a tissue-specific context.

We have previously described two supervised machine-learning approaches for prioritizing noncoding SNPs: CERENKOV2, which uses regularized gradient boosted decision trees [36] and was trained using rSNPs from the Human Gene Mutation Database (HGMD) [9], ClinVar [10], and ORegAnno [18]; and Res2s2aM [15], which uses a deep residual neural network architecture and which was trained using noncoding SNPs from the Genome-Wide Repository of Associations Between SNPs and Phenotypes, or GRASP database [12]. CERENKOV2 and Res2s2aM achieved state-of-the-art accuracy (on their respective sets of ground-truth SNPs) for discriminating rSNPs from non-functional, noncoding SNPs, but they do not include a framework for unifying the rSNP prediction score with the two key additional types of SNP-level information: SNP genotype-to-trait p-values from the GWAS, and expression QTL data. Other unified approaches that have been proposed require training sets of causal noncoding SNPs [4, 14] for the unification step, facing the user with a "chicken-and-egg problem" for applying them to a GWAS for a new trait. As we describe in the next section, in this book chapter we propose a statistical approach to unify a SNP's GWAS p-value, eQTL p-value, and rSNP prediction score, that is computationally efficient and that does not require a trait-specific training set causal noncoding SNPs.

2 Materials

Our method does not require any materials *per se*, but it requires four types of data resources, as detailed in Subheadings 3.3.2 and 3.3.4: a summary table of significant GWAS SNPs from a large representative collection of GWAS; eQTL SNP-to-gene significance test data from a large eQTL study for a tissue or tissues relevant to the GWAS trait of interest; and SNP annotation-based rSNP prediction scores for a large set of SNPs and labeled as to whether or not those SNPs are rSNPs based on external databases of experimentally validated rSNPs.

3 Methods

In this section we describe our statistical method for integrating three types of SNP-level data (genotype-to-trait p-value from the GWAS, eQTL p-value, and rSNP model prediction score) to produce a composite score suitable for ranking candidate causal noncoding SNPs within GWAS trait-associated regions. Thus, our method is intended for secondary analysis of GWAS regions of interest. We will begin by defining some notation and terminology needed to explain the method (Subheading 3.1); then we will describe the basic assumptions for our method and how these assumptions apply to the data types that our method integrates (Subheading 3.2); and then we will describe step-by-step how the three component log-likelihood ratios of our method are computed (Subheading 3.3) before concluding (Subheading 4).

3.1 Notation

Table 1 gives an overview of the notation that we use in this chapter.

3.2 Assumptions

3.2.1 GWAS

In a typical application of our method, we are searching for functional, trait-associated noncoding SNPs in summary data from a candidate-gene GWAS in which n individuals[1] have been phenotyped for a trait and genotyped at a genome-wide set of marker SNPs. Given that over 99% of SNPs in the human population are biallelic[2] [2] and given that GWAS marker SNPs are usually genotyped for only two alleles, in this chapter we will assume that we need consider only the two most common alleles for each SNP. For each SNP, we consider two hypotheses: H^c_0 denotes the null hypothesis that s is *not* functionally associated with population variation in the trait, and H^c_1 denotes the alternative hypothesis

[1] For a human GWAS, the number n of individuals in the study typically ranges from a few hundred to a few hundred thousand.

[2] For a large fraction of triallelic SNPs, the second alternative allele is rare in the population [2].

Table 1
Summary of notation conventions used in this chapter

Notation	Symbol type	Denotes
Lower-case	x	Scalar variable
Lower-case, bold	$\boldsymbol{x}$	Vector or matrix
Upper-case	X	Random variable
	$P(x)$	Shorthand for $P(X = x)$
	$O(X = x)$	Odds, $P(X = x)/(1 - P(X = x))$
	$\lambda(X = x)$	Log-odds, $\ln O(X = x)$
	$L(Y\|X)$	Likelihood, $P(X\|Y)$
	H_0, H_1	Null and alternative hypotheses
	$\mathcal{L}(H\|X)$	Log-likelihood ratio, $\ln(L(H_1\|X)/L(H_0\|X))$
	$\mathcal{F}_Z$	CDF of distribution Z

that a noncoding SNP s is a causal SNP for variation in the GWAS trait, i.e., the hypothesis that s is functionally associated with population variation in the trait. We will use the notation H^c when we want to refer to both hypotheses, as in the case of the likelihood ratio. Then for any SNP s, the germline genotypes of the n individuals in the GWAS can be represented by a vector of ternary values $\boldsymbol{g}_s \in \{0, 1, 2\}^n$ giving the counts of minor alleles for SNP s in each individual. Similarly, the phenotypes for the n individuals can be represented by a vector $\boldsymbol{d}$; in the case of a binary trait, $\boldsymbol{d} \in \{0, 1\}^n$, and in the case of a continuous trait, $\boldsymbol{d} \in (\mathbb{R}_+)^n$. Assuming that there is no significant population structure, a typical GWAS analysis for a binary trait involves computing, for each SNP s, a nominal p-value p_s^g based on the 3×2 contingency table $\boldsymbol{c}(\boldsymbol{g}_s, \boldsymbol{d})$ of genotypes and traits for the n individuals. Assuming that n is large enough that there are sufficient counts in each cell of the contingency table, and assuming there are no additional covariates, the genotype effect p-value p_s^g can be computed as

$$p_s^g = 1 - \mathcal{F}_{\chi^2_{(2)}}\left[\sum_{i=0}^{1}\sum_{j=0}^{2}\frac{(c_{ij} - \overline{c}_{ij})^2}{\overline{c}_{ij}}\right], \quad (1)$$

where $\overline{\boldsymbol{c}}$ is the matrix of expected counts based on the marginal counts of $\boldsymbol{c}(\boldsymbol{g}_s, \boldsymbol{d})$ and where $\mathcal{F}$ with subscript $\chi^2_{(\nu)}$ denotes the CDF of the χ^2 distribution with ν degrees of freedom. If there are additional covariates, logistic regression can be used to compute the single-SNP genotype effect p-value p_s^g [30]. If the trait is continuous-valued and an additive log-linear model (with coefficients α_s and β_s) is used,

$$\log d_i \sim \alpha_s + \beta_s g_{s,i} + \varepsilon_i, \tag{2}$$

then, assuming there are no other covariates, p_s^g can be computed using $\mathcal{F}_{F_{(1,n-2)}}$, the CDF of the F-distribution with parameters $(1, n - 2)$ evaluated at $f_s(g_s, \boldsymbol{d})$, the F-score for the model for s,

$$p_s^g = 1 - \mathcal{F}_{F_{(1,n-2)}}\left[f_s(g_s, \boldsymbol{d})\right]. \tag{3}$$

In a candidate-gene GWAS, for each SNP s, p_s^g is typically compared to a genome-wide significance threshold, for which a commonly used value (based on empirical analysis of the number of independent linkage disequilibrium blocks in the genotyped human genome) is 5×10^{-8}. The set of ordered pairs $\{(s, p_s^g) \mid s \in \mathbb{S}\}$ is typically called the "summary data" for the GWAS. Used by itself, this approach is relatively unbiased but omits other information about s that may inform $P(H_1^c)$ and that could increase sensitivity for detecting causal noncoding SNPs. It is useful to think of p_s^g for any given SNP as a sample from a random variable U^g; under the null hypothesis H_0^c the GWAS p-values should be uniformly distributed,

$$U^g | H^c{}_0 \sim \text{Unif}(0, 1). \tag{4}$$

3.2.2 Regulatory SNPs

In our method, we assume that a necessary, but not sufficient, condition for a noncoding SNP s to satisfy H_1^c is that it be a *regulatory SNP* (rSNP) in at least one type of tissue; an rSNP is a noncoding SNP that influences population variation in expression of a (typically vicinal but sometimes distal) gene, usually via affecting binding of a transcription factor or other chromatin-binding gene expression regulator. Let H_1^r represent the alternative hypothesis that a SNP is a *regulatory* variant in at least one tissue (which need *not* be functionally associated with the GWAS trait), and H_0^r the null hypothesis that the SNP is not a regulatory variant in *any* tissue. A fundamental assumption in our approach is that a noncoding causal SNP for the trait must be a regulatory SNP, i.e.,

$$H_1^c \Rightarrow H_1^r, \tag{5}$$

with double arrow denoting logical implication. Thus, we expect the probability that a noncoding SNP is an rSNP to be a useful prior on whether the SNP is functionally associated with population variation in a trait. There are two broad categories of information that are known to have a strong bearing on whether a noncoding SNP is an rSNP, eQTL associations and SNP annotations; our approach uses both types of information, as we describe in the next two subsections. As a consequence of Eq. 5, there are only three combinations of the H^c and H^r hypotheses that we need to consider for a given SNP s: $H_1^c \wedge H_1^r$, $H_0^c \wedge H_1^r$, and $H_0^c \wedge H_0^r$.

3.2.3 eQTL Associations

In our method we co-analyze the GWAS summary data with data from an eQTL study of a tissue type[3] that is biologically relevant to the trait, in the form of genotype data for marker SNPs[4] and genome-wide measurements of transcript abundance in the tissue type ("gene expression") in a *different group of individuals than the individuals in the GWAS*. Since our method is focused on noncoding SNPs, it is convenient to define a set $\mathbb{S}$ as the set of noncoding SNPs for which we have *both* GWAS and eQTL genotype data (which could have been obtained in part by SNP genotype imputation using haplotype data from population genome sequencing studies). We will assume that the eQTL study produces, for a given SNP s, a *cis*-eQTL p-value p^e_s for the strongest SNP-gene association (defined by minimum nominal (SNP, gene) association p-value) among all genes whose transcription start sites are within a predefined *cis*-window (for example, 1 Mbp [21]) of the SNP. We will model the distribution of such p-values, across all SNPs, as random variable U^e.

3.2.4 SNP Annotations

In general, for each SNP $s \in \mathbb{S}$, we can obtain a vector of annotations—independent of eQTL and GWAS information—that are informative for whether the SNP is a regulatory SNP or not. Such annotations can be derived from models of local phylogenetic sequence conservation, transcription factor binding site datasets, multiparameter epigenome-based annotations such as Segway or ChromHMM, and replication timing information; for an extensive discussion of SNP annotations, *see* [19, 23, 35]. We [35] and others [19, 23, 25, 26, 31, 37] have used various supervised classification algorithms to score the regulatory potential for SNPs based on a training dataset of known regulatory SNPs (i.e., noncoding SNPs for which H^r_1 holds, such as common SNPs in the Human Gene Mutation Database that are annotated as regulatory SNPs). Most of these classifiers are discriminative and thus produce a pseudo-probability score for a binary class label, which can be interpreted as $P(H^r_1 | \boldsymbol{A}_s)$, given a vector of SNP annotations $\boldsymbol{A}_s$. We assume that a specific annotation feature vector $\boldsymbol{a}_s$ has been generated for each SNP, and that a particular SNP classification algorithm (such as GWAVA [26] or CERENKOV2 [36]) has been selected that has been previously trained on a ground-truth set of known human regulatory SNPs. For example, if we were to generate $\boldsymbol{a}_s$ using CERENKOV2 [36], the SNP annotation feature vector $\boldsymbol{a}_s$ would be 258-dimensional, and for any SNP s we would obtain a CERENKOV2 rSNP prediction score $p^a_s = P(H^r_1 | \boldsymbol{a}_s)$. For an arbitrary SNP, we model the CERENKOV2 prediction score as a random variable U^a (a is short for "annotations") whose value set is the unit interval.

[3] The tissue type can range from a precisely defined primary cell type to fairly coarse-grained complex tissue types, e.g., "muscle," "heart," or "adipose tissue."

[4] For simplicity we will assume that they are imputed to the same set of SNPs as the GWAS marker SNPs.

3.3 How Our Method Works

Having outlined our method's assumptions and defined our notation, we are now able to more precisely delineate the analytical goal of our method. For a SNP s, given three outcomes:

1. $U^g = p_s^g$ from the summary data from the GWAS,
2. $U^e = p_s^e$ from an eQTL study for a tissue relevant to the GWAS trait,
3. $U^a = p_s^a$ for annotations for s,

we aim to estimate the posterior probability

$$P(H_1^c \mid p_s^g, p_s^e, p_s^a) \equiv P(H_1^c \mid U^g = p_s^g, U^e = p_s^e, U^a = p_s^a) = \frac{e^{\lambda(H_1^c \mid p_s^g, p_s^e, p_s^a)}}{1 + e^{\lambda(H_1^c \mid p_s^g, p_s^e, p_s^a)}} \tag{6}$$

by leveraging all three sources of information within a single probabilistic model. By using Bayes's theorem, the log posterior odds is the sum of the log prior odds and the log-likelihood ratio,

$$\lambda(H_1^c \mid p_s^g, p_s^e, p_s^a) = \lambda(H_1^c) + \mathcal{L}(H^c \mid p_s^g, p_s^e, p_s^a), \tag{7}$$

with the log-likelihood-ratio $\mathcal{L}(H^c \mid p_s^g, p_s^e, p_s^a)$ being the object that we will approximate with an explicit model. The prior log-odds $\lambda(H_1^c)$ is a user-definable parameter in our method. For a polygenic human trait, we would typically choose a value for $\lambda(H^c{}_1)$ in the range of -16 to -19, based on the genome-wide significance threshold of 5×10^{-8} [22]. However, for the problem of ranking candidate causal noncoding SNPs within a GWAS region of interest, a much lower log-odds ratio could be used based on the number of SNPs in the region of interest [28]; so for a noncoding region of interest with 20 SNPs, the prior log-odds for a SNP would be approximately -3 (however, to the extent that the goal is to *rank* candidate causal noncoding SNPs within a region of interest, the prior odds are invariant and irrelevant for ranking purposes.

The heart of our method is that we model the log-likelihood-ratio $\mathcal{L}(H^c \mid p_s^g, p_s^e, p_s^a)$ using a naïve Bayes-like decomposition,

$$\mathcal{L}(H^c \mid p_s^g, p_s^e, p_s^a) = \mathcal{L}_g(H^c \mid p_s^g) + \mathcal{L}_e(H^c \mid p_s^e) + \mathcal{L}_a(H^c \mid p_s^a). \tag{8}$$

This decomposition is based on the following conditional independence assumptions:

$$U^g \perp\!\!\!\perp U^e, U^a \mid H^c \tag{9}$$

$$U^a \perp\!\!\!\perp U^e \mid H^c. \tag{10}$$

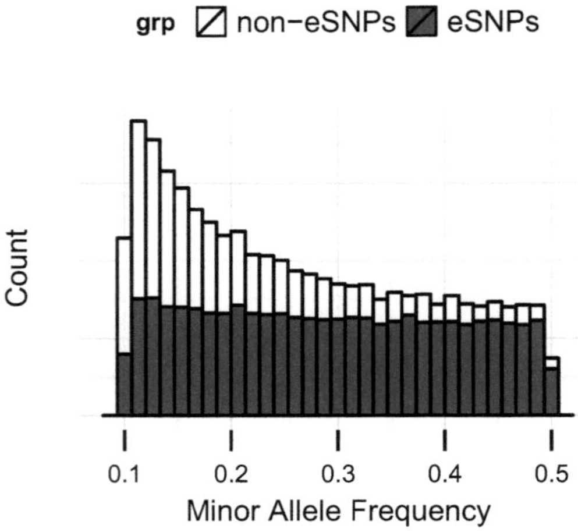

Fig. 1 Histograms of SNP minor allele frequencies, conditioned on eQTL status (eSNP or not)

From the work of Nicolae et al. [20], it seems unlikely that Eq. 9 holds *exactly;* methods development efforts to circumvent this assumption are ongoing. In any event, in our method as currently used, from a technical standpoint, Eq. 9 is important to being able to use the method on GWAS summary (as opposed to genotype-level) data. For Eq. 10, the biggest concern is the question of conditional independence of allele frequency and U^e, since lower minor allele frequency (MAF) reduces statistical power for eQTL discovery. From empirical analysis of the MAFs of significant eSNPs and non-eSNPs using data from the Genotype Tissue Expression (GTEx) project across all tissues, the violation of conditional independence appears to be modest for SNPs with MAF > 0.1 (which are the SNPs most likely to be relevant in a GWAS in any event) (Fig. 1).

3.3.1 Computing the GWAS Log-Likelihood-Ratio $\mathcal{L}_g$

For a binary trait and a biallelic SNP s, given a GWAS summary p-value p_s^g, the log-likelihood-ratio $\mathcal{L}_g(H^c \mid p_s^g)$ that corresponds to the p-value p_s^g can be approximately computed as follows:

$$\mathcal{L}_g(H^c \mid p_s^g) \simeq \frac{1}{2}\left[\mathcal{F}^{-1}_{\chi^2_{(2)}}\left(1 - p_s^g\right)\right], \tag{11}$$

assuming that the p-value was computed with no covariates other than genotype. This means that in order to use our method, it is not necessary to have access to the original genotype data for the GWAS; only summary data from the GWAS are required, provided that the asymptotic approximation underlying Wilks' theorem

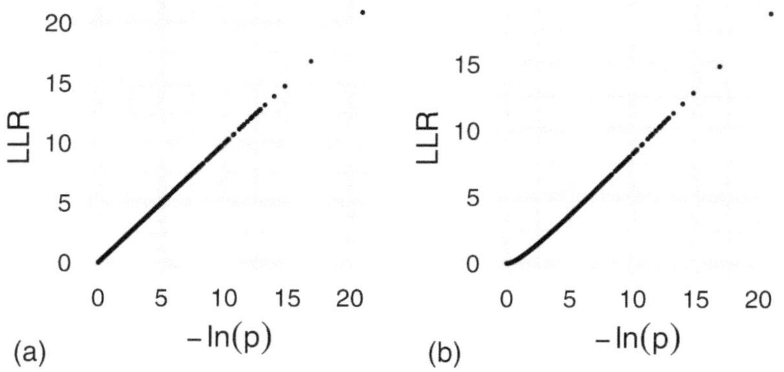

Fig. 2 The empirical relationships between $\mathcal{L}(H^c|p_s^g)$ (LLR) and $\ln(1/p_s^g)$ are asymptotically linear. (**a**) Relationship in the case of a binary trait; (**b**) relationship in the case of a continuous trait. Each mark is one of 100,000 SNPs, from a simulated GWAS of $n = 1000$ people

applies. From a practical standpoint, a key advantage of a method that relies on only summary-level data is that for many GWAS datasets, genotype data are not readily available and (depending on the dataset) may require approval from a Data Access Committee (such as for many GWAS datasets hosted on the Database of Genes and Phenotypes, dbGaP) and/or Institutional Review Board. For the case of a continuous trait, the log-likelihood ratio can be approximately computed from the p-value p_s^g by the equation

$$\mathcal{L}_g(H^c \mid p_s^g) \simeq \frac{n}{2}\ln\left[\frac{\mathcal{F}^{-1}_{F_{(1,n-2)}}(1 - p_s^g)}{n-2} + 1\right]. \tag{12}$$

In a simulated dataset of 100,000 SNP p-values from a GWAS, we see that Eq. 12 is asymptotically linear in $\ln(1/p_s)$ (Fig. 2). Thus, as the first step in our method, for each noncoding SNP s in a trait-associated GWAS region, we compute a GWAS association log-likelihood-ratio $\mathcal{L}_e(H^c|p_s^g)$ using either Eq. 11 (in the case of a binary trait) or Eq. 12 (in the case of a continuous trait). For each s, the GWAS association LLR will be combined (using Eq. 8) with two other SNP-specific LLRs $\mathcal{L}_e(H^c|p_s^e)$ and $\mathcal{L}_a(H^c|p_s^a)$, which are computed as described in Subheading 3.3.2 and Subheading 3.3.4, respectively.

3.3.2 Computing the eQTL Log-Likelihood Ratio

The eQTL log-likelihood-ratio $\mathcal{L}_e(H^c|p_s^e)$ can be computed empirically by merging data from (1) a database of GWAS trait-associated SNPs (such as GRASP) and (2) a separate eQTL study for an appropriate tissue type. For example, the Genotype Tissue Expression (GTEx) project provides tables of human eQTL association p-values for all (SNP, gene) pairs that were assayed in the project, for 48 different tissue types [5]. As an example, we computed

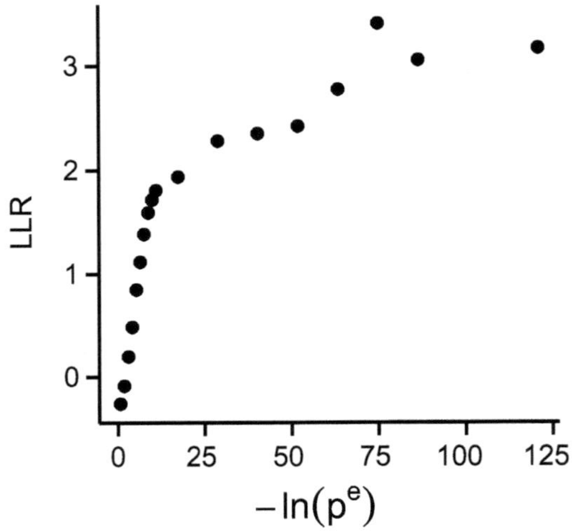

Fig. 3 Binned empirical estimates of $\mathcal{L}_e(H^c|p_s^e)$ as a function of $\ln(1/p_s^e)$. The estimates are based on data from GRASP and GTEx. Each mark indicates the log-likelihood ratio (LLR), defined by $\mathcal{L}_e(H^c|p_s^e)$, computed for all SNPs whose $\log_{10}(1/p_s^e)$ values are in a given bin (as described in the accompanying text)

$\mathcal{L}_e(H^c|p_s^e)$ using coronary artery eQTL data from GTEx and using the complete GRASP database (rel. 2.0.0.0) with a significance cutoff of $p \leq 5 \times 10^{-8}$ (Fig. 3). We used a simple bin size for $-\log_{10}p^e$ consisting of half-decade per bin between 0 and 5, and then five decades per bin between 5 and 40, and then one bin between 40 and 65. We then directly estimated the log-likelihood ratio in each bin from the empirical probability ratio. We found that p_s^e is a strong contributor to the likelihood, reaching a maximum-likelihood ratio of nearly 30 in the limit of high $\ln(1/p^e)$. For GTEx coronary artery eQTLs and GRASP GWAS SNPs, the empirical relationship between $\mathcal{L}_e(H^c|p_s^e)$ and $\ln(1/p^e)$ appears to be consistent with a bent power law with a slope of 0.24 for $p^e > 10^{-4}$ and with a slope of 0.015 for $p^e < 10^{-4}$. Thus, as the second step in our method, for each noncoding SNP s in a trait-associated GWAS region, we compute an eQTL-based log-likelihood-ratio $\mathcal{L}_e(H^c|p_s^e)$ using empirical combined analysis of datasets of genome-wide-significance GWAS SNPs and eQTL p-values from a trait-associated tissue type. For each s, the eQTL LLR will be combined (using Eq. 8) with the GWAS LLR $\mathcal{L}_g(H^c|p_s^g)$ (Subheading 3.3.1) and the SNP annotation LLR $\mathcal{L}_a(H^c|p_s^a)$ (Subheading 3.3.4).

3.3.3 Extension to Multiple Tissue Types

While in this analysis we chose to use eQTL data from a single tissue type for simplicity, our method can be readily extended to use eQTL data from multiple tissue types. In that case, we are computing $\mathcal{L}^e(H^c|\boldsymbol{p}_s^e)$, where $\boldsymbol{p}_s^e$ is a vector of p-values across the tissue types. Because eQTL data are correlated across tissues (e.g., an eSNP in coronary artery is likely to also be an eSNP in a developmentally related tissue, like aorta), it is probably not appropriate to model $\mathcal{L}^e(H^c|\boldsymbol{p}_s^e)$ using the naïve Bayes assumption. Rather, the likelihoods in $\mathcal{L}^e(H^c|\text{logit}(\boldsymbol{p}_s^e))$ can be modeled using a Gaussian Markov random field [3] which would account for non-independence of the tissue-specific p-values (where logit($\boldsymbol{p}$) means element-wise application of the logit function).

3.3.4 Computing the SNP Annotation Log-Likelihood-Ratio $\mathcal{L}_a$

From our fundamental assumption Eq. 5, it follows that the SNP annotation score-based log-likelihood-ratio $\mathcal{L}_a(H^c|p_s^a)$ can be computed in terms of two quantities, (1) a SNP annotation score-based log-likelihood ratio that the SNP is a regulatory SNP,

$$\mathcal{L}_a(H^r|p_s^a) \equiv \ln P(p_s^a|H_1^r) - \ln P(p_s^a|H_0^r), \tag{13}$$

and (2) the log-likelihood ratio that a non-causal (in the GWAS sense) SNP will be a regulatory SNP,

$$\mathcal{L}(H^r|H_0^c) \equiv \ln P(H_0^c|H_1^r) - \ln P(H_0^c|H_0^r), \tag{14}$$

by the equation

$$\mathcal{L}_a(H^c|p_s^a) = \mathcal{L}_a(H^r|p_s^a) - \ln\left(\frac{1 + e^{\mathcal{L}(H^r|H_0^c)}e^{\mathcal{L}(H^r|p_s^a)}}{1 + e^{\mathcal{L}(H^r|H_0^c)}}\right). \tag{15}$$

From a contingency table analysis of rSNPs from the Human Gene Mutation Database (HGMD, rel. 2017.2) and GWAS SNPs ($p^g < 5 \times 10^{-8}$) from GRASP, we estimate that $\mathcal{L}(H^r|H_0^c) = 0.896$. As a consequence, the variation of $\mathcal{L}_a(H^c|p_s^a)$ over the full range of logit(p_s^a) is somewhat compressed with respect to the variation of $\mathcal{L}(H^r|p_s^a)$ over the same range of logit(p_s^a), as shown in Fig. 4. In generating Fig. 4, we used the OSU18 benchmark set of SNPs [36], for which we had obtained p_s^a scores using CERENKOV2

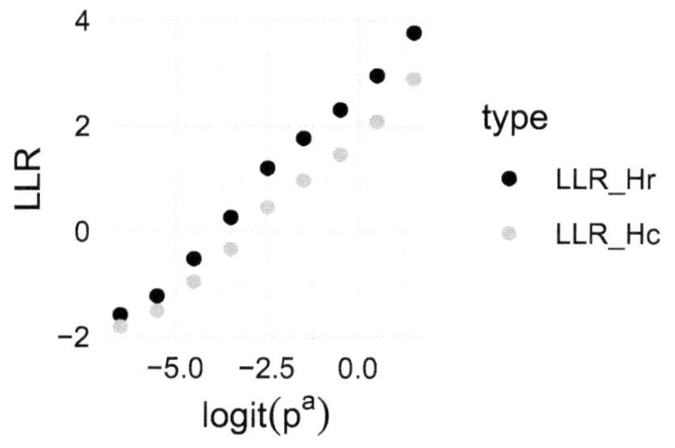

Fig. 4 Empirically evaluated log-likelihood-ratios $\mathcal{L}_a(H^c|p_s^a)$ (denoted "LLR_Hc") and $\mathcal{L}_a(H^r|p_s^a)$ (denoted "LLR_Hr") as a function of the CERENKOV2 rSNP prioritization score, p_s^a. The relationship between $\mathcal{L}_a(H^c|p_s^a)$ and logit(p_s^a) is approximately linear with a slope of 0.7

[36]. With an empirical estimate of the relationship between p_s^a and the log-likelihood $\mathcal{L}_a(H^c|p_s^a)$ now in hand, for any noncoding SNP *s*, assuming the SNP's annotation-based rSNP score p_s^a can be obtained (which requires only a trained rSNP prediction model and the model's required feature vector $\boldsymbol{a}_s$), we can compute $\mathcal{L}_a(H^c|p_s^a)$. This completes the third required term in Eq. 8. While we have illustrated this procedure using SNP annotation-based rSNP prediction scores from CERENKOV2, the same procedure would be applicable using any rSNP prediction model that generates probabilistic prediction scores, for example, GWAVA [26], RSVP [23], or DeepSEA [37]. Naturally, the empirical relationship between $\mathcal{L}_a(H^c|p_s^a)$ and $\text{logit}(p_s^a)$ will be tool-dependent and would need to be re-estimated if rSNP scores from a different tool are used.

4 Notes

In this book chapter we have described a practical and statistically integrated method for combining eQTL data and rSNP prediction scores with GWAS summary *p*-values in order to identify candidate causal noncoding SNPs in trait-associated regions. The method requires no training using example SNPs (in contrast to other unified methods as described in Subheading 1.1); this is a key advantage since there is a paucity of human noncoding SNPs that are *validated* causal for organism-level trait variation. Furthermore, the equation for integrating the three types of SNP-level data (Eq. 8) is compatible with vectorized computational implementation and thus we anticipate it would easily scale to application for secondary analysis of loci from thousands of GWAS. A possible extension of the method would be to incorporate the explained trait variance (based on a single-SNP model) as an additional data type to be integrated in the log-likelihood model, though the significant differences in polygenicity of traits may pose a challenge to deriving a single likelihood model. Moreover, a key area for future methods development is to try to extend our likelihood-based approach to incorporate multi-SNP fine-mapping information, where it is available for a given GWAS dataset. The importance of future continued development of computational data fusion approaches for ranking candidate causal noncoding SNPs in GWAS regions of interest is underscored by the rapid growth of GWAS datasets worldwide and the rapid growth of functional genomic and molecular cellular datasets that can complement GWAS datasets.

Acknowledgements

This work was supported by the National Science Foundation (award numbers 1557605-DMS and 1553728-DBI to S.A.R.), the PhRMA Foundation (Informatics Grant to S.A.R.), and the Oregon State University Division of Health Sciences (Interdisciplinary Research Grant Award to S.A.R.).

References

1. Bryzgalov LO, Antontseva EV, Matveeva MY, Shilov AG, Kashina EV, Mordvinov VA, Merkulova TI (2013) Detection of regulatory SNPs in human genome using ChIP-seq ENCODE data. PLoS One 8(10):e78833
2. Cao M, Shi J, Wang J, Hong J, Cui B, Ning G (2015) Analysis of human triallelic SNPs by next-generation sequencing. Ann Hum Genet 79(4):275–281
3. Chen M, Cho J, Zhao H (2011) Incorporating biological pathways via a Markov random field model in genome-wide association studies. PLOS Genet 7(4):e1001353
4. Gao L, Uzun Y, Gao P, He B, Ma X, Wang J, Han S, Tan K (2018) Identifying noncoding risk variants using disease-relevant gene regulatory networks. Nat Commun 9(1):702
5. GTEx Consortium (2015) Human genomics. The genotype-tissue expression (GTEx) pilot analysis: multitissue gene regulation in humans. Science 348(6235):648–660
6. Gulko B, Hubisz MJ, Gronau I, Siepel A (2015) A method for calculating probabilities of fitness consequences for point mutations across the human genome. Nat Genet 47 (3):276–283
7. Ionita-Laza I, McCallum K, Xu B, Buxbaum JD (2016) A spectral approach integrating functional genomic annotations for coding and noncoding variants. Nat Genet 48 (2):214–220
8. Kircher M, Witten DM, Jain P, O'Roak BJ, Cooper GM, Shendure J (2014) A general framework for estimating the relative pathogenicity of human genetic variants. Nat Genet 46 (3):310–315
9. Krawczak M, Cooper DN (1997) The human gene mutation database. Trends Genet 13 (3):121–122
10. Landrum MJ, Lee JM, Riley GR, Jang W, Rubinstein WS, Church DM, Maglott DR (2014) ClinVar: public archive of relationships among sequence variation and human phenotype. Nucleic Acids Res 42(D1):D980–D985
11. Lee D, Gorkin DU, Baker M, Strober BJ, Asoni AL, McCallion AS, Beer MA (2015) A method to predict the impact of regulatory variants from DNA sequence. Nature Genet 47(8):955–961, gkm-SVM
12. Leslie R, O'Donnell CJ, Johnson AD (2014) GRASP: analysis of genotype-phenotype results from 1390 genome-wide association studies and corresponding open access database. Bioinformatics 30(12):i185–i194
13. Li MJ, Wang LY, Xia Z, Sham PC, Wang J (2013) GWAS3D: detecting human regulatory variants by integrative analysis of genome-wide associations, chromosome interactions and histone modifications. Nucleic Acids Res 41(W1): W150–W158
14. Li MJ, Pan Z, Liu Z, Wu J, Wang P, Zhu Y, Xu F, Xia Z, Sham PC, Kocher JPA, Li M, Liu JS, Wang J (2016) Predicting regulatory variants with composite statistic. Bioinformatics 32(18):2729–2736
15. Liu Z, Yao Y, Wei Q, Weeder B, Ramsey SA (2019) Res2s2aM: deep residual network-based model for identifying functional noncoding SNPs in trait-associated regions. In: Liu Z (ed) Proceedings of the 24th Pacific symposium on biocomputing
16. Macintyre G, Bailey J, Haviv I, Kowalczyk A (2010) is-rSNP: a novel technique for *in silico* regulatory SNP detection. Bioinformatics 26 (18):i524–i530
17. Maurano MT, Humbert R, Rynes E, Thurman RE, Haugen E, Wang H, Reynolds AP, Sandstrom R, Qu H, Brody J *et al* (2012) Systematic localization of common disease-associated variation in regulatory DNA. Science 337(6099):1190–1195
18. Montgomery SB, Griffith OL, Sleumer MC, Bergman CM, Bilenky M, Pleasance ED, Prychyna Y, Zhang X, Jones SJM (2006) ORegAnno: an open access database and curation system for literature-derived promoters, transcription factor binding sites and regulatory variation. Bioinformatics 22(5):637–640

19. Montgomery SB, Griffith OL, Schuetz JM, Brooks-Wilson A, Jones SJM (2007) A survey of genomic properties for the detection of regulatory polymorphisms. PLOS Comput Biol 3 (6):e106
20. Nicolae DL, Gamazon E, Zhang W, Duan S, Dolan ME, Cox NJ (2010) Trait-associated SNPs are more likely to be eQTLs: annotation to enhance discovery from GWAS. PLOS Genet 6(4):e1000888
21. Ongen H, Buil A, Brown AA, Dermitzakis ET, Delaneau O (2016) Fast and efficient QTL mapper for thousands of molecular phenotypes. Bioinformatics 32(10):1479–1485
22. Panagiotou OA, Ioannidis JPA, Genome-Wide Significance Project (2012) What should the genome-wide significance threshold be? Empirical replication of borderline genetic associations. Int J Epidemiol 41(1):273–286
23. Peterson TA, Mort M, Cooper DN, Radivojac P, Kann MG, Mooney SD (2016) Regulatory single-nucleotide variant predictor increases predictive performance of functional regulatory variants. Hum Mutat 37 (11):1137–1143
24. Quang D, Xie X (2016) DanQ: a hybrid convolutional and recurrent deep neural network for quantifying the function of DNA sequences. Nucleic Acids Res 44(11):e107
25. Quang D, Chen Y, Xie X (2015) DANN: a deep learning approach for annotating the pathogenicity of genetic variants. Bioinformatics 31(5):761–763
26. Ritchie GRS, Dunham I, Zeggini E, Flicek P (2014) Functional annotation of noncoding sequence variants. Nat Methods 11 (3):294–296
27. Riva A (2012) Large-scale computational identification of regulatory SNPs with rSNP-MAPPER. BMC Genet 13(Suppl 4):S7
28. Schaid DJ, Chen W, Larson NB (2018) From genome-wide associations to candidate causal variants by statistical fine-mapping. Nat Rev Genet 19(8):491
29. Schaub MA, Boyle AP, Kundaje A, Batzoglou S, Snyder M (2012) Linking disease associations with regulatory information in the human genome. Genome Res 22 (9):1748–1759
30. Stranger BE, Stahl EA, Raj T (2011) Progress and promise of genome-wide association studies for human complex trait genetics. Genetics 187(2):367–383
31. Torkamani A, Schork NJ (2008) Predicting functional regulatory polymorphisms. Bioinformatics 24(16):1787–1792
32. Welter D, MacArthur J, Morales J, Burdett T, Hall P, Junkins H, Klemm A, Flicek P, Manolio T, Hindorff L, Parkinson H (2014) The NHGRI GWAS Catalog, a curated resource of SNP-trait associations. Nucleic Acids Res 42(D1):D1001–D1006. Accessed in 2016
33. Xiao R, Scott LJ (2011) Detection of *cis*-acting regulatory SNPs using allelic expression data. Genetic Epidemiol 35(6):515–525
34. Xu H, Gregory SG, Hauser ER, Stenger JE, Pericak-Vance MA, Vance JM, Züchner S, Hauser MA (2005) SNPselector: a web tool for selecting SNPs for genetic association studies. Bioinformatics 21(22):4181–4186
35. Yao Y, Liu Z, Singh S, Wei Q, Ramsey SA (2017) CERENKOV: computational elucidation of the regulatory noncoding variome. In: Proceedings of the 8th ACM international conference on bioinformatics, computational biology, and health informatics. ACM, New York, pp 79–88
36. Yao Y, Liu Z, Wei Q, Ramsey SA (2019) CERENKOV2: improved detection of functional noncoding SNPs using data-space geometric features. BMC Bioinform 20:63 https://doi.org/10.1186/s12859-019-2637-4
37. Zhou J, Troyanskaya OG (2015) Predicting effects of noncoding variants with deep learning-based sequence model. Nature Methods 12(10):931–934

Chapter 7

Statistical and Machine Learning Methods for eQTL Analysis

Junjie Chen and Conor Nodzak

Abstract

An immense amount of observable diversity exists for all traits and across global populations. In the post-genomic era, equipped with efficient sequencing capabilities and better genotyping methods, we are now able to more fully appreciate how regulation of gene expression is consequential to one's genotypes in coding and non-coding DNA. The identification of genetic loci that contribute to quantifiable variation in genetic expression is critical in further improving our understanding of the biological regulation of complex traits. Expression quantitative traits loci (eQTLs) mapping studies have provided a powerful suite of techniques for genome wide analysis to detect these regulatory effects. However, a typical eQTL analysis relies on a large number of samples with many genetic variants to achieve robust power and significance for detection. With this in mind, eQTL analysis brings about distinct computational and statistical challenges that require advanced methodological development to overcome. In recent years, many statistical and machine learning methods for eQTL analysis have been developed with the ability to provide a more complex perspective towards the identification of relationships between genetic variation and genetic expression. In this chapter, we provide a comprehensive review of statistical and machine learning methods. We will present various machine learning methods based upon regularization terms and several other statistical analysis methods. Finally, we will discuss prior knowledge integration and hyperparameter optimization.

Key words eQTL analysis, Regularization term, Multi-task learning, Statistical analysis

1 Introduction

To understand how the genetic variation affects gene expression is one of the fundamental tasks in computational biology [1]. The most common genetic variation is single nucleotide polymorphisms (SNPs) that may cause phenotype variation such as diseases or different gene expression traits [2]. Although the association mapping of causal expression Quantitative Trait Loci (eQTLs) remains challenging as the variation of complex traits is a result of contributions of many genetic variations, eQTL analysis has have been conducted by many studies [3–5] and shows a powerful way to extend our knowledge on the associations among genetic variants and gene expressions. Specifically, eQTL analysis treats gene

Xinghua Mindy Shi (ed.), *eQTL Analysis: Methods and Protocols*, Methods in Molecular Biology, vol. 2082,
https://doi.org/10.1007/978-1-0716-0026-9_7,

expression as quantitative phenotypes, and intends to find genetic variants that are significantly associated with changes in gene expression. These identified associations can help reveal biological mechanism processes underlying living systems, discover genetic factors, and pathways that cause disease and phenotype variation [6].

In recent years, high-throughput technologies including microarrays and sequencing have enabled the identification of genetic variation and the quantification of gene expression at whole genome level. The immense number of genetic variants and that of genes results in the search space is tremendous. For example, there could be millions of SNPs and over twenty thousand genes in a genome wide eQTL analysis in humans [4]. With such large scale of data and the goal of understanding complex relationship between gene expression and genetic variants, eQTL analysis brings about distinct computational and statistical challenges.

Various feature selection methods widely used in the machine learning community have recently been applied in eQTL analysis. A general illustration of machine learning model for eQTL analysis is as shown in Fig. 1. In eQTL analysis, a set of samples from a population or cohort is assessed for their genotypes of genetic variants, followed by gene expression profiling of the same individuals. The genotypes can be called by microarrays or DNA sequencing based methods, while gene expression profiles can be quantified by gene expression arrays or mRNA sequencing. In general, the eQTL analysis can be formalized into a linear regression model in Eq. 1.

$$\mathbf{Y} = \mathbf{XB} + \mathbf{E} \tag{1}$$

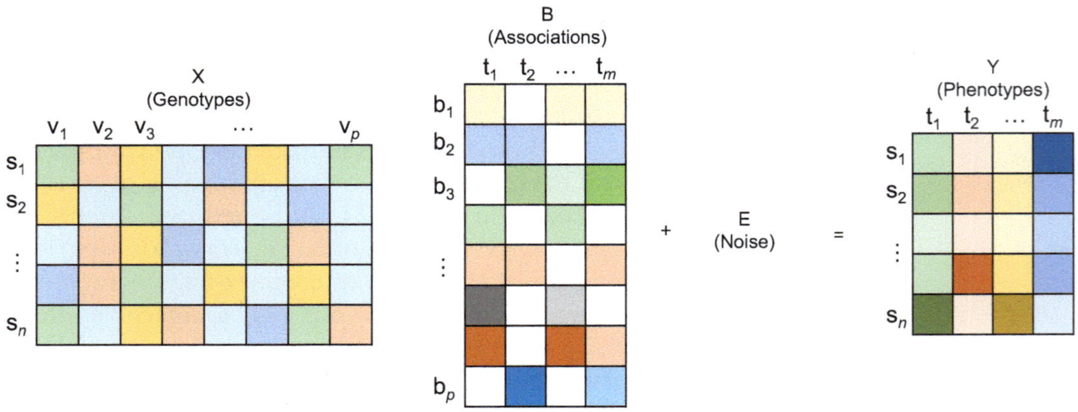

Fig. 1 An illustration of machine learning models for eQTL analysis. Standard eQTL analysis involves a direct association test between markers of genetic variation with gene expression levels typically measured in tens or hundreds of individuals

where **X** and **Y** are the input data, representing the genotypes of genetic variants and gene expression traits respectively. Note that eQTL analysis can consider one genotype or multiple genotypes in one regression. Most of the studies analyze the multiple related genotypes at the same time, because they share useful information and multiple related tasks helps to improve the generalized performance of all the tasks. The objective of such a linear model is to find a coefficient matrix **B** that would convert the genotypes matrix **X** to gene expression traits matrix **Y**, with some white-noise term **E**. Specifically, **X** is an $n \times p$ matrix, where n is number of individuals, each individual contains p genetic variants. If there is one gene expression trait, **Y** is a vector with n observations. If there are multiple gene expression traits, **Y** is an $n \times m$ matrix, where n is number of individuals, and each individual contains m quantitative gene expression traits. $\mathbf{B}_{p \times m}$ is the association coefficient matrix denoting the connection strengths between traits and genetic variants, and **E** is a Gaussian white-noise term with constant variance σ^2. In this view, genetic variants are treated as features and gene expressions are viewed as labels. In eQTL analysis, both the feature matrix on genetic variants and the label matrix on gene expressions are usually high-dimensional. Besides that, the number of features (i.e., genetic variants) and the number of labels (i.e., gene expressions) are significantly larger than the number of samples. Sparse learning methods are widely used in eQTL analysis. Regularization term based machine learning methods works well with analyzing high-dimensional and large-scale genomic data in that they can use penalized terms to impose sparsity to simultaneously enable scalable feature selection and dimension reduction to avoid overfitting. Such methods including classical Least Absolute Shrinkage and Selection Operator (Lasso) [7], Elastic Net [8], and their variations are suitable and scalable for analyzing genomic data.

In this chapter, we first review Lasso and Elastic Net methods and their variations in single task. Then, we introduce the multitask learning methods for eQTL analysis. At last, we further discuss the importance of hyperparameter tuning in machine learning model building, and present commonly employed optimization procedures for their configuration.

2 eQTL Analysis Using Regularization Term Based Machine Learning Methods

In eQTL analysis, genomic data is typically high-dimensional, where the number of genetic variants is significantly larger than sample size. Sparse learning methods are widely used in eQTL analysis, because sparse learning methods use various regularization terms to impose sparsity, shrinkage some coefficients of features to zero, as a way to avoid overfitting. The method of sparse modeling is to find the optimal coefficient matrix **B**, by minimizing the square loss function L plus a regularization term Ω in Eq. 2.

$$\min_{\mathbf{B}} L(\mathbf{B}) + \Omega(\mathbf{B}) \tag{2}$$

These methods include classical Lasso [7], Elastic Net [8], and some variants from Lasso and Elastic Net. We first review these methods in single eQTL analysis task. Then we introduce the multi-task learning for eQTL analysis.

2.1 Lasso and Its Variations

Lasso is a regression-based method that performs variable selection with ℓ^1 norm regularization in order to avoid the overfitting and enhance interpretability of the statistical model. Since it was originally introduced in 1996 [7], a number of regularization terms in Laso have been created to remedy certain limitations of the original technique and to make the method more useful for particular problems. Almost all of these extended methods focus on respecting or utilizing different types of dependencies among the covariates.

2.1.1 Lasso

Lasso minimizes the residual sum of squares subject to the sum of the absolute value of the coefficients being less than a constant. Then the general objective of Lasso is Eq. 3

$$\min_{\beta} \frac{1}{2}\|y - \mathbf{X}\beta\|_2^2 + \lambda\|\beta\|_1 \tag{3}$$

where $\mathbf{X}$ is the genotype matrix and β is coefficient vector; y is the gene expression vector. Lasso model uses regularization term ℓ^1 norm which shrinks the parameters of the majority of the genetic variants to zero, and those variants corresponding non-zero terms are selected as the identified eQTL associations.

However, the Lasso model assumes that genetic variants are independent and the expression of genes is not correlated. As a result, it will inevitably miss many complexes. Yet observed, cases where multiple genetic variants jointly affect the co-expressions of multiple genes. To account for relatedness of different genes or traits, Lasso models are extended to various models detailed below.

2.1.2 Group Lasso

The aim of the group Lasso method [9] is to allow predefined groups of covariates to be selected into or out of a model together, so that all the members of a particular group are either included or not included. Group Lasso is more natural than Lasso for eQTL analysis because genes and proteins often participate in known pathways. This approach may be useful to an investigator more interested in which pathways are related to an outcome than the individual genes themselves. The objective function for the group Lasso is a generalization of the standard Lasso objective Eq. 4

$$\min_{\beta} \frac{1}{2} \left\| y - \sum_{j=1}^{J} \mathbf{X}_j \beta_j \right\|_2^2 + \lambda \sum_{j=1}^{J} \sqrt{p_j} \|\beta_j\|_2^2 \tag{4}$$

where the genomic matrix $\mathbf{X}$ and coefficient vector β have been replaced by a collection of submatrices X_j and corresponding coefficient vectors β_j, one for each of the J groups, p_j is the length of β_j. Additionally, the regularization term is now a sum over ℓ^2 norms. If the size of each group is 1, then this reduces to the standard Lasso, while if there is only a single group, it reduces to ridge regression. Since the penalty reduces to an ℓ^2 norm on the subspaces defined by each group, it cannot select out only some of the covariates from a group, just as ridge regression cannot. However, because the penalty is the sum over the different subspace norms, as in the standard Lasso, the constraint has some non-differential points, which correspond to some subspaces being identically zero. Therefore, it can set the coefficient vectors corresponding to some subspaces to zero, while only shrinking others.

However, the non-overlapping group structure in group Lasso limits its applicability in practice. Overlapping group Lasso [10, 11] allows covariates to be shared between different groups with an assumption that a gene variant were to occur in two genes expression.

2.1.3 Sparse Group Lasso

Sometimes we would like to induce sparsity both intra and inter groups [12, 13] extend the group Lasso to sparse group Lasso, which can select individual covariates within a group by adding an additional ℓ^1 penalty to each group subspace.

$$\min_{\beta} \frac{1}{2} \left\| y - \sum_{j=1}^{J} \mathbf{X}_j \beta_j \right\|_2^2 + (1-\alpha)\lambda \sum_{j=1}^{J} \sqrt{p_j} \|\beta_j\|_2^2 + \alpha\lambda \|\beta\|_1 \tag{5}$$

where $\alpha \in [0, 1]$. α is a convex combination of Lasso and group Lasso penalties. $\alpha = 0$ gives the group Lasso, $\alpha = 1$ gives the standard Lasso.

2.1.4 Fused Lasso

Genetic variants often have linkage disequilibrium (LD) [14], some adjacent gene variants may have important spatial or temporal structure that must be accounted for during analysis. These genetic variants with strong LD may have concordant affects on the expression of a gene, thus they have similar coefficient values. Fused Lasso [15] was introduced to extend the use of Lasso to approach this exact type of problem. The fused Lasso objective function is Eq. 6.

$$\min_{\beta} \frac{1}{2} \|y - \mathbf{X}\beta\|_2^2 + (1-\alpha)\lambda \|\beta\|_1 + \alpha \sum_{i=1}^{p} |\beta_i - \beta_{i-1}| \tag{6}$$

where p is the number of genetic variants. The first constraint is just the typical Lasso regularization term, but the second constraint directly penalizes large changes with respect to the temporal or spatial structure, which forces the coefficients to vary in a smooth fashion that reflects the underlying logic of the system being studied.

2.1.5 Clustered Lasso

Clustered Lasso [16] is a generalization of fused Lasso that identifies and groups relevant covariates based on their non-zero coefficients. The basic idea is to penalize the differences between the coefficients so that non-zero ones make clusters together. This can be modeled in Eq. 7.

$$\min_{\beta} \frac{1}{2}\|y - \mathbf{X}\beta\|_2^2 + (1-\alpha)\lambda\|\beta\|_1 + \alpha\sum_{i<j}^{p}|\beta_i - \beta_j| \tag{7}$$

In contrast, one can first cluster variables into highly correlated groups, and then extract a single representative covariates from each cluster [17].

2.2 Elastic Net Based Method and Its Variations

Lasso provides the sparse solution for feature selection where the number of covariates is greater than the sample size ($p > n$). However, Lasso can select only n covariates (even when more covariates are associated with the outcome) and it tends to select only one random covariate from any set of highly correlated covariates. Additionally, even when $n > p$, if the covariates are strongly correlated, ridge regression [18] tends to perform better than Lasso. Thus, Elastic Net [8] was introduced to solve the limitation of Lasso and ridge regression.

2.2.1 Elastic Net

Compared with Lasso, Elastic Net adds an additional ridge regression-like penalty term which improves performance when the number of genetic variants is significantly larger than the sample size. As a result, Elastic Net can select strongly correlated variables together, and improves overall prediction accuracy. The Elastic Net extends Lasso by adding an additional ℓ^2 penalty term, as Eq. 8 shown.

$$\min_{\beta}\|y - X\beta\|_2^2 + (1-\alpha)\lambda\|\beta\|_1 + \alpha\lambda\|\beta\|_2^2 \tag{8}$$

where $\alpha \in [0, 1]$ is convex combination between Lasso and ridge penalties. $\alpha = 0$ provides Lasso, and $\alpha = 1$ provides ridge regression.

2.2.2 Group Elastic Net

Similar with group Lasso [9], group Elastic Net is to encourage sparsity at the group level. This ℓ^1 norm regularization encourages sparsity at the group level where an entire β_j might become 0. The squared ℓ^2 norm regularization is in similar spirit to Elastic net, and addresses some of the issues of Lasso. The objective formula is shown in Eq. 9.

$$\min_{\beta} \frac{1}{2}\left\| y - \sum_{j=1}^{J} \mathbf{X}_j\beta_j \right\|_2^2 + \sum_{j=1}^{J} \sqrt{p_j}\lambda((1-\alpha)\|\beta_j\|_1 + \alpha\|\beta_j\|_2^2) \tag{9}$$

where the genomic matrix $\mathbf{X}$ and coefficient vector β have been replaced by a collection of submatrices X_j and corresponding coefficient vectors β_j. There are J groups in total, and p_j is the length of β_j. $\alpha \in [0, 1]$ is convex combination between Lasso and ridge penalties. Comparing with sparse group Lasso that adds ℓ^1 norm to the whole space, group Elastic Net adds ℓ^1 norm to each group space. If the size of each group is 1, then this reduces to the standard Elastic Net.

2.2.3 Sparse Group Elastic Net

While group sparsity can help assess certain data structures, it is desirable in many instances to also capture elementwise sparsity. Comparing with sparse group Lasso that has been conducted in the term of ℓ^2/ℓ^1 penalized regression, sparse group Elastic Net uses an $\ell^\infty/\ell^1/\ell^2$-based penalty [19]. The ℓ^∞ norm induces group sparsity in the presence of noisy data. The objective formula is in Eq. 10.

$$\min_{\beta} \frac{1}{2}\left\| y - \sum_{j=1}^{J} \mathbf{X}_j\beta_j \right\|_2^2 + \sum_{j=1}^{J} \sqrt{p_j}(\lambda_1\|\beta_j\|_\infty + \lambda_2\|\beta_j\|_1 + \lambda_3\|\beta_j\|_2^2) \tag{10}$$

where λ_1, λ_2, and λ_3 are non-negative tuning parameters controlling the degree of group sparsity, within group sparsity, and collinearity, respectively. Note, when $\lambda_1 = 0$ this is the group Elastic Net problem; setting $\lambda_1 = 0$ and $\lambda_2 = 0$ is the ridge regression; setting $\lambda_1 = 0$ and $\lambda_3 = 0$ is the group Lasso.

2.3 Multi-Task Learning Method

The eQTL hotspots are always associated with multiple complex gene expression traits, which are in the same pathway and regulated at the same time. If we perform eQTL analysis for each gene expression trait, we will miss the inherent complex relationship among multiple gene expression traits. Multi-task learning [20–22] (MTL), contrasted with single-task learning methods above, solves multiple traits associations at the same time, while leveraging shared information contained in multiple related tasks to help improve the general performance of all the tasks.

If there are multiple gene expression $\mathbf{Y}_{n\times m}$, where n is number of individuals, and each individual contains m quantitative gene expression traits. $\mathbf{B}_{p\times m}$ is the association coefficient matrix denoting the connection strengths between traits and genetic variants. MTL optimizes the coefficient matrix, $\mathbf{B}$, by minimizing the loss function L plus a regularization term Ω in Eq. 11.

$$\min_{\mathbf{B}} \frac{1}{2}\|\mathbf{Y} - \mathbf{XB}\|_F^2 + \lambda\Omega \tag{11}$$

where the $\|A\|_F = \sqrt{\sum_{i=1}^{m}\sum_{j=1}^{n}|a_{ij}|^2}$ is Frobenius norm.

In an ideal situation, all tasks are related, thus information can be shared across all tasks. Mean-regularized MTL, joint feature learning, alternating structural optimization(ASO), and the shared Parameter Gaussian Process can be used for this situation. However, there are outliers in some situations in which several tasks are not related to the others. Many multi-task learning variations were proposed for solving different assumptions, like the dirty model and robust MTL.

2.3.1 Mean-Regularized MTL

This method assumes that all tasks are similar and their coefficients are very close. This implies that all coefficients are "close" to some mean function. The objective function is Eq. 12.

$$\min_{\mathbf{B}} \frac{1}{2}\|\mathbf{Y} - \mathbf{X}\mathbf{B}\|_F^2 + \lambda \sum_{i=1}^{m} \left\| \mathbf{B}_i - \frac{1}{m}\sum_{s=1}^{m} \mathbf{B}_s \right\|_2^2 \tag{12}$$

where m is the number of gene expression. The penalty term is the sum of deviation of each task from the mean. This penalty enforces all task coefficient vectors towards their mean that is controlled by λ, in result of their coefficients are very close.

2.3.2 Joint Feature Learning

Although multi-task learning uses the shared information in multiple tasks to improve performance, sparse learning is still important for high-dimensional problems. Joint feature learning [23], also called block-sparse multiple regression, is proposed to impose sparsity over all of the genetic variants and gene expression traits. Its objective function is formulated by Eq. 13.

$$\min_{\mathbf{B}} \frac{1}{2}\|\mathbf{Y} - \mathbf{X}\mathbf{B}\|_F^2 + \lambda\|\mathbf{B}\|_{2,1} \tag{13}$$

where $\|\mathbf{B}\|_{2,1} = \sum_{i=1}^{p} \|\beta_i\|_2$ is ℓ^1/ℓ^2 norm, also called block-sparse regularization, p is the number of genetic variations. While in the single-task setting, the ℓ^1 norm is employed to shrink the parameter vector β. For MTL, ℓ^2 norm is first added across each row β_i containing the parameter corresponding to the ith feature across all tasks. Then ℓ^1 norm of this vector forces all but a few entries of β_i to be 0. As shown in Fig. 2, some blocks are shrunk to zeros. In general, any $\|\mathbf{B}\|_{q,1}$ regularization can be used to constraint the parameters, when $q > 1$ introduces the sparsity.

2.3.3 Dirty Model

The ℓ^1/ℓ^q norm block-regularizations used in joint feature learning [23] establishes strong guarantees on recovery, even under high-dimensional scaling. However, authors [23] also cautioned that the performance of such block-regularized methods are very dependent on the extent to which the features are shared across tasks. If the extent of overlap is less than a given threshold, or even if parameter values in the shared features are highly uneven, then

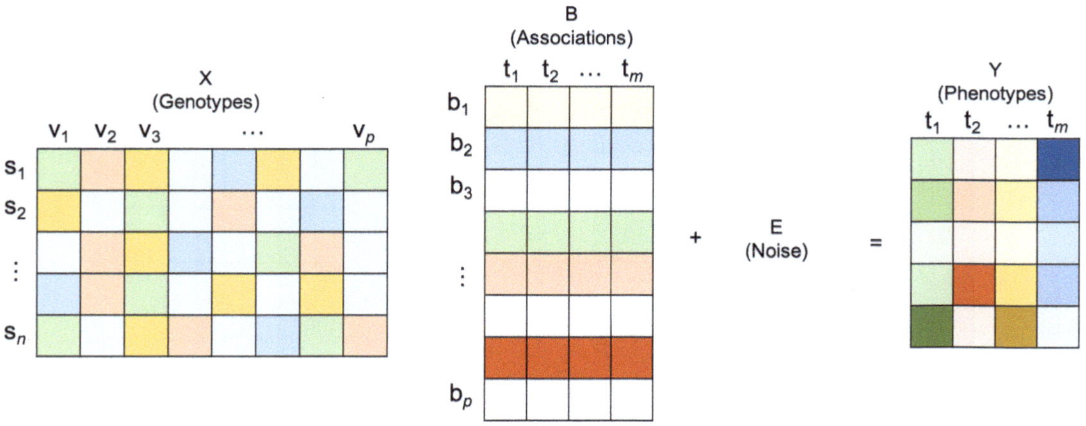

Fig. 2 An illustration of joint feature learning MTL. Joint feature learning MTL uses block-square regularization which forces some row blocks to zeros

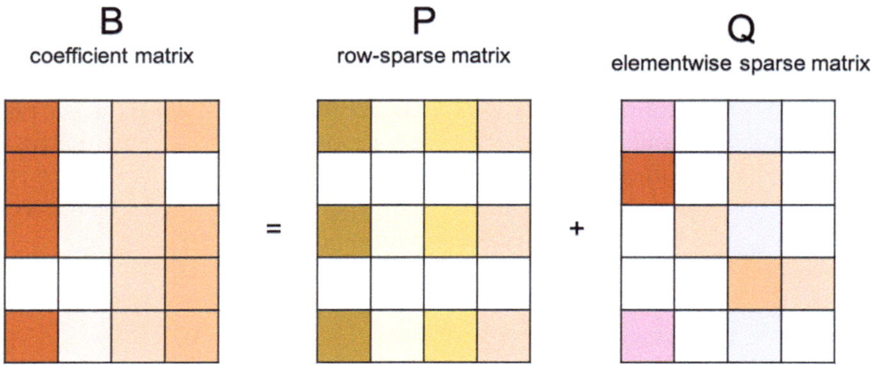

Fig. 3 An illustration of dirty model MTL. Dirty model uses two regularization terms, block-regularized penalty for those related task and ℓ^1 regularization for elementwise sparse matrix

block ℓ^1/ℓ^q regularization could actually perform worse than simple separate elementwise ℓ^1 regularization.

It is often too strict to assume the tasks have same structure. Thus the dirty model [24] is proposed to model such "dirty data", which may not fall into single neat structural bracket. The main idea is that while any one structure might not capture the data, a superposition of structural classes might. The parameter matrix is decomposed into a row-sparse matrix (corresponding to the overlapping or shared features) and an elementwise sparse matrix (corresponding to the non-shared features), as shown in Fig. 3.

$$\min_{\mathbf{B}} \frac{1}{2}\|Y - \mathbf{X}(\mathbf{P}+\mathbf{Q})\|_F^2 + \lambda_1\|\mathbf{P}\|_{1,q} + \lambda_2\|\mathbf{Q}\|_{1,1} \tag{14}$$

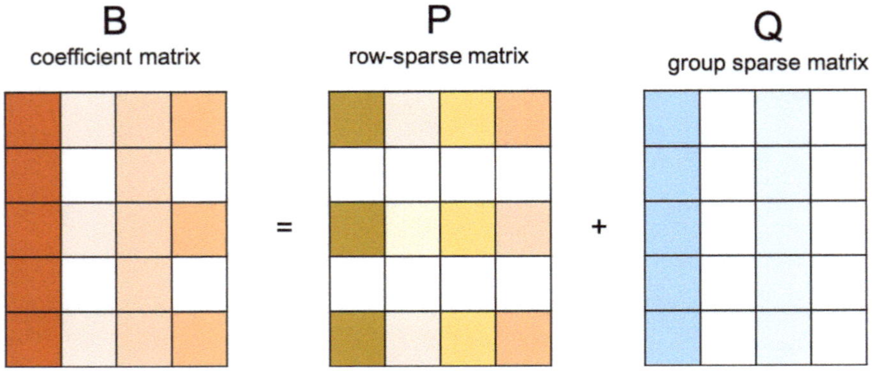

Fig. 4 An illustration of robust MTL. Robust MTL decomposes the coefficient matrix as two sub-matrices and penalizes them through row-block regularization and column-block regularization, respectively

where **P** is row-sparse matrix which is penalized by ℓ^1/ℓ^q norm block regularization, and **Q** is elementwise sparse matrix which is penalized by ℓ^1 regularization.

2.3.4 Robust MTL

Robust MTL [25] is another model for solving outlier tasks. It is a similar strategy that decomposed the coefficient matrix into two sub-matrices. The difference is that Robust MTL uses row-sparse regularization and column-sparse regularization terms, as shown in Fig. 4. Its objective function is Eq. 15.

$$\min_{\mathbf{B}} \frac{1}{2}\|Y - \mathbf{X}(\mathbf{P}+\mathbf{Q})\|_F^2 + \lambda_1\|\mathbf{P}\|_{1,q} + \lambda_2\left\|\mathbf{Q}^{\mathrm{T}}\right\|_{1,q} \tag{15}$$

where **P** is row-sparse matrix and **Q** is column-sparse matrix. Both regularization terms are ℓ^1/ℓ^q norm. The difference between first and second penalty terms is that **Q** is transformed. $\|\mathbf{P}\|_{1,q}$ can jointly select sparse features. $\|\mathbf{Q}^{\mathrm{T}}\|_{1,q}$ can jointly select the outlier tasks.

2.4 Multi-Task Learning for Complicated Tasks

eQTL analysis aims to identify the genetic loci associated with gene expression and the penalized regression models, with a proper penalty, presented in the above sections are often suitable for high-dimensional biological data. However, detecting eQTLs remains a challenge due to complex underlying mechanisms and the very large number of genetic loci involved compared to the number of samples. Thus, to address this issue, it is desirable to take advantage of the structure of the data and prior information about genomic locations such as conservation scores, transcription factor binding sites, biological knowledge of gene expression networks, and linkage disequilibrium (LD) structure between loci in a high-noise background. Depending on the structure of task relatedness, information can be shared in selective tasks. For this situation, clustered MTL, tree MTL, and Graph MTL are proposed.

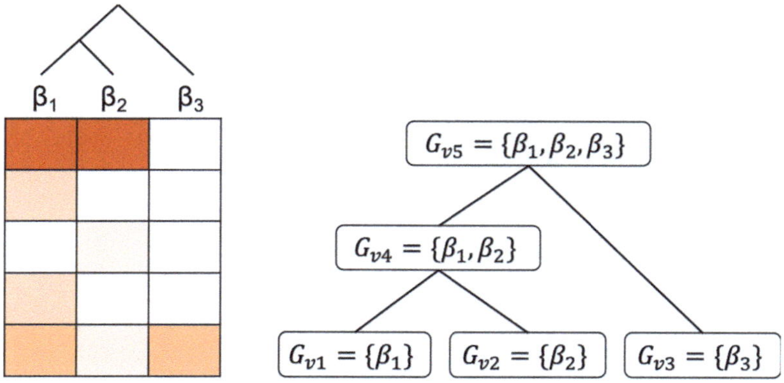

Fig. 5 An illustration of tree-guided multi-task learning. The hierarchical clustering tree represents the correlation structure in responses. The first two responses are highly correlated according to the clustering tree, and are likely to be influenced by the same covariates. Groups of variables associated with each node of the tree in tree-Lasso penalty

2.4.1 Tree Multi-task Learning

Several extensions of the multi-task learning model have been proposed to take consideration of the network structure underlying the relatedness of genes. A tree-guided model [26] is developed based on the idea that co-expressed genes share a larger common set of genetic variants comparing those independent genes, as Fig. 5 shown. In this model, the tree structure can be obtained using a hierarchical clustering tree [27] on labels provided by the user.

In tree-guided multi-task learning, the tasks should be equipped with tree structures, in which tasks that belong to a same node are similar to each other and the similarity between two tasks is structured and related to the depth of the ancestor node. The objective function is Eq. 16.

$$\min_{\beta} \frac{1}{2} \|\mathbf{Y} - \mathbf{XB}\|_F^2 + \lambda \sum_{j}^{J} \sum_{v \in V} \sqrt{p_j} \left\| \beta_j^{G_v} \right\|_2^2 \tag{16}$$

where $\beta_j^{G_v}$ is a vector of regression coefficients $\{\beta_j^k | k \in G_v\}$, J are groups in total, and p_j is the length of β_j.

Note that this tree-guided multi-task learning model is an extension of a group Lasso model. Additionally, an adaptive multi-task Lasso model [26] considers the correlation among gene expressions, while incorporating the priors on SNPs such as regulatory features for these SNPs.

2.4.2 Single Graph-Guided Multitask Learning

Graph-guided multi-task model [28] can also be used to estimate genetic variants that perturb a subset of highly correlated genes. Let $G = (V, E)$ be a weighted graph that represents the relatedness of the genes, where V is the set of vertices and E is the set of edges. Genes are vertices and edges represent the relatedness between two

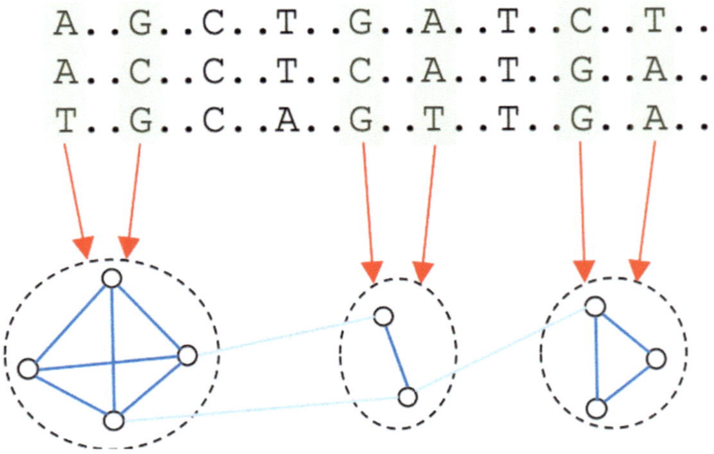

Fig. 6 Illustrations for association analysis with graph-guided MTL. This method considers quantitative traits networks with edge weights

genes in G. The idea here is that if two genes are correlated and an genetic variant is associated with one of these two genes, then the probability that this variant is associated with the other gene is greater, as Fig. 6 shown.

As seen in Eq. 17, a regularization term is added to reflect the contribution from correlated genes, with a structure dictated by a graph G.

$$min_{\mathbf{B}}\|\mathbf{Y}-\mathbf{XB}\|_2^2+\lambda\|\mathbf{B}\|_1+ \gamma\sum_{e_{m,l}\in E} w(e_{m,l})\sum_{j=1}^{J}\left|\beta_{jm}-sign(r_{m,l})\beta_{jl}\right| \tag{17}$$

where, $w(e_{m,l})$ is a weight assigned to the edge $e_{m,l}$ in graph G and $r_{m,l}$ is the correlation between y^m and y^l. λ and γ are regularization parameters that control the complexity of the model. A larger value for γ leads to a greater fusion effect.

Such a graph-guided multi-task model in Eq. 17 can learn the associations between one particular genetic variant and a group of correlated genes. The associations among genetic variants and genes will be reflected by **B** matrix. The paper [28] only considers $w(e,l) = |r|$, but any positive, monotonically increasing function of the absolute value of correlations can be used. The sign$(r_{m,l})$ weights the fusion penalty for each edge such that β_{jm} and β_{jl} for highly correlated outputs with large $|r_{m,l}|$ receives a greater fusion effect than other pairs of outputs with weaker correlations. The sign $(r_{m,l})$ indicates that two negatively correlated outputs are encouraged to have the same set of relevant covariates with the same absolute value of regression coefficients of the opposite sign.

The regularization term in Eq. 17 is closely related to that in fused Lasso [15]. Thus, a graph-guided multi-task Lasso model can be viewed as a generalization of the fused Lasso model in that fusion is dictated by the topology of input graphs, rather than physical proximity.

2.4.3 Two-Graph-Guided Multi-Task Model

The graph-guided multi-task model [28] can be further extended by incorporating the correlation among genetic variants. The rationale is that if one genetic variant is associated with the expression of a specific gene, then another genetic variant, which co-occurs frequently with the first genetic variant, would be more likely to be associated with the expression of this particular gene, as shown in Fig. 7. This rationale can be formalized as adding another regularization term to Eq. 17, which leads to a two-graph guided multi-task Lasso model as proposed in [29].

The two-graph guided multi-task model [29] can be learned by minimizing the objective function in Eq. 18.

$$
\begin{aligned}
& min_{\mathbf{B}} \|\mathbf{Y} - \mathbf{XB}\|_F^2 + \lambda \|\mathbf{B}\|_1 + \\
& \gamma_1 \sum_{e_{m,l} \in E_1} w(e_{m,l}) \sum_{j=1}^{J} |\beta_{jm} - sign(r_{m,l})\beta_{jl}| + \\
& \gamma_2 \sum_{e_{f,g} \in E_w} w(e_{f,g}) \sum_{k=1}^{K} |\beta_{fk} - sign(r_{f,g})\beta_{gk}|
\end{aligned}
\tag{18}
$$

where, $w(e_{m,l})$ and $r_{m,l}$ are a weight assigned to the edge $e_{m,l}$ and correlation between y^m and y^l in graph G. $w(e_{f,g})$ and $r_{f,g}$ are a weight assigned to the edge $e_{f,g}$ and correlation between y^f and y^g in graph G.

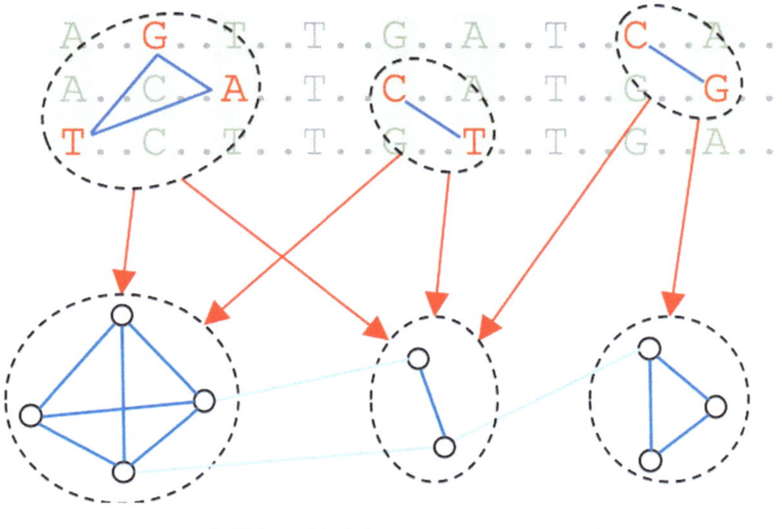

Fig. 7 Illustrations of MTL guided by genetic variants subnetwork and the quantitative traits subnetwork

The two-graph guided multi-task model captures the scenario that multiple genetic variants, by forming a subnetwork, can affect the expression of multiple correlated genes. Such a model has a nice sparse property that it allows flexible structured sparsity both on genetic variants and genes. The two-graph guided multi-task models can be seen as a generalization of several multi-task feature selection methods proposed before [15, 26, 28, 30].

2.4.4 Jointly Structured Input and Output Model

Another model, named jointly structured input and output model [31], has been developed to consider the problem of learning a multi-task regression model while taking advantage of the prior information on structures on both the inputs (genetic variations) and outputs (expression levels). The model implements an l_1/l_2 regularized multi-task regression [32]. The structured input-output Lasso model can be achieved by solving the following optimization problem.

$$min_{\mathbf{B}}\|\mathbf{Y}-\mathbf{XB}\|_F^2+\lambda_1\|\mathbf{B}\|_1+\gamma_1\sum_{k=1}^{K}\sum_{g\in V_1}\|\beta_k^g\|_2+\gamma_2\sum_{j=1}^{J}\sum_{h\in V_2}\left\|\beta_k^h\right\|_2 \quad (19)$$

The last two terms in Eq. 19 reflect the contribution from the groups, denoted as V_1, of genetic variants, and the contribution from the groups of genes, defined as V_2, respectively.

3 Hyperparameter Optimization

Machine learning methods are being applied to a wide range of problems in biology. These methods often rely on configuring high level parameters, or hyperparameters, such as regularization hyperparameters in sparse learning models like λ in Lasso methods. The importance of hyperparameter selection becomes evident when one considers that different hyperparameters will lead to different results. In simple terms, a hyperparameter is used in the process of determining the parameters used by the model. Different machine learning models utilize hyperparameters in various ways, such as the ℓ^1/ℓ^2 norm in regularized regression methods imposes penalty on model complexity. There are several different ways to tune them in order to increase model performance including manual tuning, grid search, random search, and global methods such as Bayesian optimization. For eQTL studies, the goal is to find optimal hyperparameters that can be evaluated by maximization of the R^2 score.

General manual search of hyperparameters involves the sequential testing of hyperparameters by an expert user, whereas more recently, grid search and random search optimization methods have been widely used in the machine learning community and continue to be popular choices due to the simplicity of their implementations. Grid search utilizes a uniform input grid of possible hyperparameters and exhaustively search for the optimal values using all possible combinations. This approach allows one to have a reliable estimate of hyperparameter selection in low dimensions, but can be considered wasteful due to computational expenditures in hyperparameters with poor performance. In high-dimensional spaces, however, the performance of hyperparameters may be influenced asymmetrically by each dimension meaning that some hyperparameters do not matter, which leads researchers to devise the random search method. Random search attempts to optimize hyperparameters by random selection of values over a nonuniform search space to produce a more efficient manner around uninformative areas. This performance contrasts grid search in the ability to test more distinct values over the same number of trials due to the grid's uniformity [33]. The increase in the speed of the random search hyperparameter optimization method is achieved by testing few values in the search space and finding good parameters that maximize the marginal likelihood based on a Normal distribution for positive real values, such as eQTL studies [34]. Bayesian optimization uses a null prior distribution and is therefore a model-based approach to hyperparameter tuning. In order to find the hyperparameters that perform well, the Bayesian optimization approach uses information learned previously to update future iterations to map a probability to each hyperparameter. In doing so, this method attempts to resolve the highest probability hyperparameters and has been shown that this method can produce better results than grid and random search [35]. One important consideration when performing hyperparameter optimization is overfitting, in which the model performs very well on training data yet does poorly on new data it has not seen before. While regularized regression can help avoid this problem, a popular choice to mitigate this problem in machine learning is to implement cross-validation, to tune hyperparameters based on the averages of models built over multiple subsets of the training data.

4 Discussion

In this paper, we reviewed various machine learning methods that are based on regularization terms for eQTL analysis. Due to the fact that the number of genetic variations is much larger than observations, feature selection methods are widely applied to identify eQTLs. Lasso is one of classical feature selection methods, which shrinks the parameters of unrelated genetic variants to zero, and those related variants are selected as eQTLs. However, the classical Lasso model assumes independence among genetic variants and cannot account for biological features like linkage disequilibrium between genetic variants. To account for dependencies between different genetic variants, various Lasso models were proposed, such as group Lasso, sparse group Lasso, fused Lasso, and clustered Lasso. Although Lasso provides the sparse solution for feature selection where the number of covariates is greater than the sample size ($p > n$), it can select only n covariates even when more covariates are associated with the outcome, and it tends to select only one covariate from any set of highly correlated covariates. Elastic Net was introduced to solve the limitation of Lasso and ridge regression. Similarly with Lasso, many variants of Elastic Net were proposed to account for the relatedness of different genetic variants, such as group Elastic Net and Sparse Group Elastic Net. Common eQTL hotspots are often associated with multiple complex gene expression traits, and tend to be found in the same biochemical pathways with similar regulation patterns. Multi-task learning solves for multiple concurrent trait associations, while leveraging shared information contained in multiple related tasks to help improve the generalization performance of all the tasks. Such methods are enhanced by leveraging prior knowledge that was either represented as gene pairs, SNP pairs, gene networks, or genetic interaction networks, and may be modeled using a tree/graph-guided multi-task method, which hierarchically organized tree/graph structure based on relatedness among quantitative traits.

Although those methods guided with prior knowledge are appealing, the implementation is obviously limited by domain knowledge and available genomic information about the species. In the future more reference genomes and genetic data will continually be produced so it has become more urgent to propose novel eQTL mapping method that can incorporate a priori knowledge. Another problem faced by eQTL research methods that employ machine learning based methods is hyperparameter optimization. The performance of eQTL analysis often relies on configuring many hyperparameters, such as the regularization terms for feature selection. Manual tuning, grid search, random search, and global methods such as Bayesian optimization provide several different ways to tune hyperparameters in order to increase model performance. However, for a typical eQTL analysis, there are millions of genetic variants meaning that the hyperparameter tuning procedures are very time consuming operations.

Traditional and commonly used methods for eQTL analysis employ simple linear regression, which is limited in terms of its ability to glean associations from real processes of gene expression. For eQTL analysis, the genetic variants could be thought as points in genetic regulatory networks, which regulate the gene expressions by interactions, or edges in these genetic regulatory networks. Therefore, Graph Neural Networks (GNNs) [36] could potentially be utilized to reveal complex relationships between genetic variants and gene expression in an eQTL analysis. GNNs are connectionist models that capture the dependence of graphs via message passing between the nodes of the graph. In recent years, systems based on Graph Convolutional Network (GCN) and Gated Graph Neural Network (GGNN) have demonstrated ground-breaking performance on many tasks, like learning molecular fingerprints [37, 38], and predicting protein interfaces [39]. As we continue to employ more complex models in eQTL research, we may be able to better capture the ways in which variants contribute to gene expression and other quantitative traits. Knowledge-based and graph-based machine learning methods for eQTL analysis allow better utilize known biological interdependence between genes as well as capture complex epistatic interactions between variants to provide a more robust contextualization and understanding of variation.

References

1. Rockman MV, Kruglyak L (2006) Genetics of global gene expression. Nat Rev Genet 7 (11):862
2. Yang J, Benyamin B, McEvoy BP, Gordon S, Henders AK, Nyholt DR, Madden PA, Heath AC, Martin NG, Montgomery GW *et al* (2010) Common snps explain a large proportion of the heritability for human height. Nat Genet 42(7):565
3. Cookson W, Liang L, Abecasis G, Moffatt M, Lathrop M (2009) Mapping complex disease traits with global gene expression. Nat Rev Genet 10(3):184
4. Cheung VG, Spielman RS (2009) Genetics of human gene expression: mapping DNA variants that influence gene expression. Natl Rev Genet 10(9):595
5. Stranger BE, Forrest MS, Clark AG, Minichiello MJ, Deutsch S, Lyle R, Hunt S, Kahl B, Antonarakis SE, Tavaré S *et al* (2005) Genome-wide associations of gene expression variation in humans. PLoS Genet 1(6):e78
6. Stranger BE, Forrest MS, Dunning M, Ingle CE, Beazley C, Thorne N, Redon R, Bird CP, De Grassi A, Lee C *et al* (2007) Relative impact of nucleotide and copy number variation on gene expression phenotypes. Science 315 (5813):848–853
7. Tibshirani R (1996) Regression shrinkage and selection via the lasso. J R Stat Soc Ser B (Methodol) 58(1):267–288
8. Zou H, Hastie T (2005) Regularization and variable selection via the elastic net. J R Stat Soc Ser B (Stat Methodol) 67(2):301–320
9. Yuan M, Lin Y (2006) Model selection and estimation in regression with grouped variables. J R Stat Soc Ser B (Stat Methodol) 68(1):49–67
10. Jacob L, Obozinski G, Vert JP (2009) Group lasso with overlap and graph lasso. In: Proceedings of the 26th annual international conference on machine learning. ACM, New York, pp 433–440
11. Yuan L, Liu J, Ye J (2011) Efficient methods for overlapping group lasso. In: Advances in neural information processing systems, pp 352–360
12. Simon N, Friedman J, Hastie T, Tibshirani R (2013) A sparse-group lasso. J Comput Graph Stat 22(2):231–245
13. Friedman J, Hastie T, Tibshirani R (2010) A note on the group lasso and a sparse group lasso. arXiv preprint arXiv:10010736
14. Reich DE, Cargill M, Bolk S, Ireland J, Sabeti PC, Richter DJ, Lavery T, Kouyoumjian R, Farhadian SF, Ward R *et al* (2001) Linkage disequilibrium in the human genome. Nature 411(6834):199
15. Tibshirani R, Saunders M, Rosset S, Zhu J, Knight K (2005) Sparsity and smoothness via the fused lasso. J R Stat Soc Ser B (Stat Methodol) 67(1):91–108
16. She Y *et al* (2010) Sparse regression with exact clustering. Electron J Stat 4:1055–1096
17. Reid S, Tibshirani R (2016) Sparse regression and marginal testing using cluster prototypes. Biostatistics 17(2):364–376
18. Hoerl AE, Kennard RW (1970) Ridge regression: biased estimation for nonorthogonal problems. Technometrics 12(1):55–67
19. Samarov DV, Allen D, Hwang J, Lee YJ, Litorja M (2017) A coordinate-descent-based

approach to solving the sparse group elastic net. Technometrics 59(4):437–445

20. Argyriou A, Evgeniou T, Pontil M (2007) Multi-task feature learning. In: Advances in neural information processing systems, pp 41–48
21. Evgeniou T, Pontil M (2004) Regularized multi-task learning. In: Proceedings of the tenth ACM SIGKDD international conference on knowledge discovery and data mining. ACM, New York, pp 109–117
22. Zhang Y, Yang Q (2017) A survey on multi-task learning. arXiv preprint arXiv:170708114
23. Negahban S, Wainwright MJ (2008) Joint support recovery under high-dimensional scaling: benefits and perils of ℓ^1, ∞-regularization. In: Proceedings of the 21st international conference on neural information processing systems. Curran Associates, Red Hook, pp 1161–1168
24. Jalali A, Sanghavi S, Ruan C, Ravikumar PK (2010) A dirty model for multi-task learning. In: Advances in neural information processing systems, pp 964–972
25. Chen J, Zhou J, Ye J (2011) Integrating low-rank and group-sparse structures for robust multi-task learning. In: Proceedings of the 17th ACM SIGKDD international conference on knowledge discovery and data mining. ACM, New York, pp 42–50
26. Kim S, Xing EP *et al* (2012) Tree-guided group lasso for multi-response regression with structured sparsity, with an application to eQTL mapping. Ann Appl Stat 6 (3):1095–1117
27. Murtagh F (1983) A survey of recent advances in hierarchical clustering algorithms. Comput J 26(4):354–359
28. Kim S, Xing EP (2009) Statistical estimation of correlated genome associations to a quantitative trait network. PLoS Genet 5(8):e1000587
29. Chen X, Shi X, Xu X, Wang Z, Mills R, Lee C, Xu J (2012) A two-graph guided multi-task lasso approach for eQTL mapping. In: Artificial intelligence and statistics, pp 208–217
30. Lee S, Zhu J, Xing EP (2010) Adaptive multi-task lasso: with application to eQTL detection. In: Advances in neural information processing systems, pp 1306–1314
31. Lee S, Xing EP (2012) Leveraging input and output structures for joint mapping of epistatic and marginal eQTLS. Bioinformatics 28(12): i137–i146
32. Obozinski G, Taskar B, Jordan M (2007) Joint covariate selection for grouped classification. Technical Report, Statistics Department, UC Berkeley
33. Bergstra J, Bengio Y (2012) Random search for hyper-parameter optimization. J Mach Learn Res 13:281–305
34. Varma S, Das S (2018) Deep learning. https://srdas.github.io/DLBook/HyperParameterSelection.html#tuning-hyper-parameters
35. Bergstra J, Yamins D, Cox DD (2013) Making a science of model search: hyperparameter optimization in hundreds of dimensions for vision architectures. J Mach Learn Res 28:I-115–I-123
36. Zhou J, Cui G, Zhang Z, Yang C, Liu Z, Sun M (2018) Graph neural networks: a review of methods and applications. arXiv preprint arXiv:181208434
37. You J, Liu B, Ying Z, Pande V, Leskovec J (2018) Graph convolutional policy network for goal-directed molecular graph generation. In: Advances in neural information processing systems, pp 6410–6421
38. De Cao N, Kipf T (2018) Molgan: an implicit generative model for small molecular graphs. arXiv preprint arXiv:180511973
39. Fout A, Byrd J, Shariat B, Ben-Hur A (2017) Protein interface prediction using graph convolutional networks. In: Advances in neural information processing systems, pp 6530–6539

Chapter 8

Sparse Regression Models for Unraveling Group and Individual Associations in eQTL Mapping

Wei Cheng, Xiang Zhang, and Wei Wang

Abstract

As a promising tool for dissecting the genetic basis of common diseases, expression quantitative trait loci (eQTL) study has attracted increasing research interest. Traditional eQTL methods focus on testing the associations between individual single-nucleotide polymorphisms (SNPs) and gene expression traits. A major drawback of this approach is that it cannot model the joint effect of a set of SNPs on a set of genes, which may correspond to biological pathways. To alleviate this limitation, in this chapter, we propose *geQTL*, a sparse regression method that can detect both group-wise and individual associations between SNPs and expression traits. *geQTL* can also correct the effects of potential confounders. Our method employs computationally efficient technique, thus it is able to fulfill large scale studies. Moreover, our method can automatically infer the proper number of group-wise associations. We perform extensive experiments on both simulated datasets and yeast datasets to demonstrate the effectiveness and efficiency of the proposed method. The results show that *geQTL* can effectively detect both individual and group-wise signals and outperform the state-of-the-arts by a large margin. This book chapter well illustrates that decoupling individual and group-wise associations for association mapping is able to improve eQTL mapping accuracy, and inferring individual and group-wise associations.

Key words eQTL mapping, Group-wise association, Computation efficiency

1 Introduction

Expression quantitative trait loci (eQTL) mapping aims at identifying single-nucleotide polymorphisms (SNPs) that influence the expression level of genes. It has been widely applied to analyze the genetic basis of gene expression and molecular mechanisms underlying complex traits [2, 19]. In a typical eQTL study, the association between each expression trait and each SNP is assessed separately [9, 25, 29]. This approach does not consider the interactions among SNPs and among genes. However, multiple SNPs may interact with each other and jointly influence the phenotypes [14]. This assumption will inevitably miss complex cases where multiple genetic variants jointly affect the co-expressions of multiple genes. It has been observed in biological experiments that the

Xinghua Mindy Shi (ed.), *eQTL Analysis: Methods and Protocols*, Methods in Molecular Biology, vol. 2082,
https://doi.org/10.1007/978-1-0716-0026-9_8, © Springer Science+Business Media, LLC, part of Springer Nature 2020

joint effect of multiple SNPs to a phenotype may be non-additive [14], and genes from the same biological pathway are usually co-regulated [21] by the same genetic basis. The biological process contains both individual effects and joint effects between SNPs and genes [20]. A straightforward approach to detect associations between sets of SNPs and a gene expression level can be done using the standard gene set enrichment analysis [11]. Wu et al. [26] further proposed the variance component models for SNP set testing. Braun et al. employed aggregation-based approaches to cluster SNPs [3]. In [16], Listgarten et al. further considered the potential confounding factors.

However, there are two limitations for these approaches. First, these methods typically only consider SNPs from pre-defined pathways or gene ontology categories, which are far from being complete. Second, these methods can only detect the mapping of SNP set and a single gene expression level. To better elucidate the genetic basis of gene expression, it is a crucial challenge to understand how multiple modestly-associated SNPs interact to influence the group of genes [14]. In this chapter, we refer to this kind of eQTL mapping to find associations between group of SNPs and group of gene expression levels as the *group-wise* eQTL mapping. An example is shown in Fig. 1. Note that an ideal model should allow overlaps between SNP sets and between gene sets, that is, a SNP or gene may participate in multiple individual and group-wise associations [14]. In literature, *group-wise* eQTL mapping has attracted increasing research interest recently. For example, Xu et al. [6] proposed a two-graph-guided multi-task Lasso approach to infer group-wise eQTL mapping. However, it required the grouping information of both SNPs and genes available as prior knowledge, which may not be practical for many applications. Besides, it is not able to correct the effects of confounding factors.

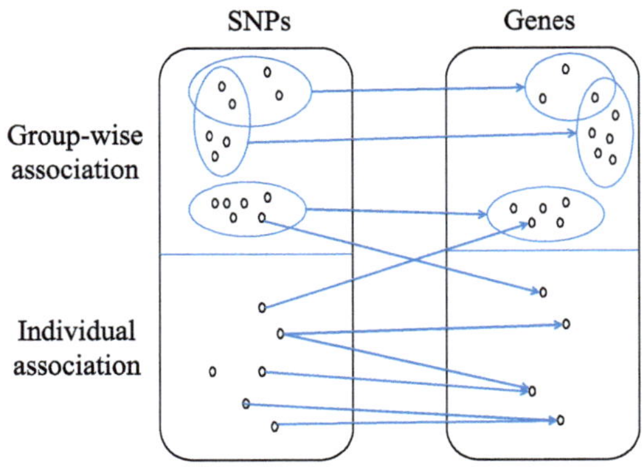

Fig. 1 An illustration of individual and group-wise associations. Ellipses represent the groups of SNPs and genes. Blue arrows between SNPs and genes represent identified associations

In this chapter, we propose a novel method, *geQTL*, to automatically detect individual and group-wise associations in eQTL studies. It uses a two-layer feature selection strategy and adopts efficient optimization techniques, which make it suitable for large scale studies. Moreover, geQTL can automatically infer the optimal number of group-wise associations. We perform extensive experiments on both simulated datasets and yeast datasets to demonstrate the effectiveness and efficiency of the proposed method.

2 The Proposed Approach

2.1 Preliminaries

Important notations used in this chapter are listed in Table 1. In this chapter, for each sample, the data of SNPs and genes are denoted by column vectors. Let $\mathbf{x} = [x_1, x_2, \ldots, x_K]^T$ denote the K SNPs. Here, $x_i \in \{0, 1, 2\}$ denotes a random variable corresponding to the i-th SNP (For example, 0, 1, 2 may encode the homozygous major allele, heterozygous allele, and homozygous minor allele, respectively.). Let $\mathbf{z} = [z_1, z_2, \ldots, z_N]^T$ denote the N genes in the study. z_j denotes a continuous random variable corresponding to the j-th gene expression. Let $\mathbf{X} = \{\mathbf{x}_h | 1 \leq h \leq H\} \in \mathbb{R}^{K \times H}$ be the SNP matrix. We use $\mathbf{Z} = \{\mathbf{z}_h | 1 \leq h \leq$

Table 1
Notations

Symbols	Description
K	Number of SNPs
N	Number of genes
H	Number of samples
M	Number of group-wise associations
$\mathbf{x}$	Random variables of K SNPs
$\mathbf{z}$	Random variables of N genes
$\mathbf{y}$	Latent variables to model group-wise associaiton
$\mathbf{X} \in^{K \times H}$	SNP matrix data
$\mathbf{Z} \in^{N \times H}$	Gene expression matrix data
$\mathbf{A} \in^{M \times K}$	Group-wise association coefficient matrix between $\mathbf{x}$ and $\mathbf{y}$
$\mathbf{B} \in^{N \times M}$	Group-wise association coefficient matrix between $\mathbf{y}$ and $\mathbf{z}$
$\mathbf{C} \in^{N \times K}$	Individual association coefficient matrix between $\mathbf{x}$ and $\mathbf{y}$
$\alpha, \beta, \gamma, \rho$	Regularization parameters
$\mathbf{R} \in^{N \times K}$	Indicator matrix showing which elements in $\mathbf{C}$ can be nonzero

$H\} \in \mathbb{R}^{N \times H}$ to denote the matrix of gene expression levels. H denotes the number of samples in consideration.

The traditional linear regression model for association mapping between $\mathbf{x}$ and $\mathbf{z}$ is

$$\mathbf{z} = \mathbf{W}\mathbf{x} + \boldsymbol{\mu} + \boldsymbol{\varepsilon}, \tag{1}$$

where $\mathbf{z}$ is a linear function of $\mathbf{x}$ with coefficient matrix $\mathbf{W}$, $\boldsymbol{\mu}$ is an $N \times 1$ translation factor vector. And $\boldsymbol{\varepsilon}$ is the additive noise of Gaussian distribution with zero mean and variance $\gamma\mathbf{I}$, where γ is a scalar. That is, $\boldsymbol{\varepsilon} \sim \mathcal{N}(\mathbf{0}, \gamma\mathbf{I})$.

In association studies, sparsity is a reasonable assumption because only a small fraction of genetic variants are expected to be associated with a set of gene expression traits. This can be modeled as a feature selection problem. For example, the standard Lasso [25] can be used in association mapping, which applies ℓ_1 penalty on $\mathbf{W}$ for sparsity.

If both $\mathbf{X}$ and $\mathbf{Z}$ are standardized, the objective function of Lasso is formulated as

$$\min_{\mathbf{W}} ||\mathbf{Z} - \mathbf{W}\mathbf{X}||_F^2 + \eta||\mathbf{W}||_1, \tag{2}$$

where $||\cdot||_F$ denotes the Frobenius norm, $||\cdot||_1$ is the ℓ_1-norm. η is the empirical parameter for the ℓ_1 penalty. $\boldsymbol{W}$ is the parameter (also called weight) matrix parameterizing the space of linear functions mapping from $\mathbf{X}$ to $\mathbf{Z}$.

Confounding factors, such as unobserved covariates, experimental artifacts, and unknown environmental perturbations, may mask real signals and lead to spurious findings. LORS [27] uses a low-rank matrix $\mathbf{L} \in \mathbb{R}^{N \times H}$ to account for the variations caused by hidden factors. The objective function of LORS is

$$\min_{\mathbf{W}, \mathbf{L}} ||\mathbf{Z} - \mathbf{W}\mathbf{X} - \mathbf{L}||_F^2 + \eta||\mathbf{W}||_1 + \rho||\mathbf{L}||_*, \tag{3}$$

where $||\cdot||_*$ is the nuclear norm [27]. ρ is the regularization parameter to control the rank of $\mathbf{L}$. $\mathbf{L}$ is a low-rank matrix assuming that there are only a small number of hidden factors influencing the gene expression levels.

When we fix $\{\mathbf{W}\}$, we can optimize $\{\mathbf{L}\}$ by using singular value decomposition (SVD) according to the following lemma.

Lemma 1 ([17]): Suppose that matrix $\mathbf{O}$ *has rank r. The solution to the optimization problem*

$$\min_{\mathbf{S}} \frac{1}{2} ||\mathbf{O} - \mathbf{S}||_F^2 + \lambda||\mathbf{S}||_* \tag{4}$$

is given by $\hat{\mathbf{S}} = \mathbf{H}_\lambda(\mathbf{O})$, *where* $\mathbf{H}_\lambda(\mathbf{O}) = \mathbf{U}\mathbf{D}_\lambda\mathbf{V}^{\mathrm{T}}$ *with* $\mathbf{D}_\lambda = \mathrm{diag}[(d_1 - \lambda)_+, \ldots, (d_r - \lambda)_+]$, $\mathbf{U}\mathbf{D}\mathbf{V}^{\mathrm{T}}$*is the singular value decomposition (SVD) of* $\mathbf{O}$, $\mathbf{D} = \text{diag}[d_1, \ldots, d_r]$, *and* $(d_i - \lambda)_+ = \max((d_i - \lambda), 0), (1 \leq i \leq r)$.

Thus, for fixed $\mathbf{W}$, the formula for updating $\mathbf{L}$ is

$$\mathbf{L} \leftarrow \mathbf{H}_\lambda(\mathbf{Z} - \mathbf{WX}). \tag{5}$$

Both Lasso and LORS do not consider the existence of group-wise associations. Below, we will introduce the proposed model to infer both group-wise and individual associations for eQTL mapping.

2.2 geQTL

In geQTL, individual associations between SNPs and genes are modeled by following the Lasso-based strategy. Group-wise associations are inferred using a two-layer feature selection method. Since multiple SNPs may have joint effect on a group of genes, and such effect may be accomplished through complex biological processes, we introduce latent variables to bridge sets of SNPs and sets of genes. Specifically, we assume that there exist latent factors regulating the gene expression level, which serve as bridges between the SNPs and the genes. The latent variables are denoted by $\mathbf{y} = [y_1, y_2, \ldots, y_M]^{\mathrm{T}}$. Here, M ($M \ll \min(K, N)$) is the total number of latent variables representing group-wise associations. The relationship between $\mathbf{x}$ and $\mathbf{y}$ can be represented as

$$\mathbf{y} = \mathbf{Ax} + \boldsymbol{\varepsilon}_1, \tag{6}$$

where

$$\boldsymbol{\varepsilon}_1 \sim \mathcal{N}(\mathbf{0}, \sigma_1^2 \mathbf{I}_M).$$

$\mathbf{A} \in \mathbb{R}^{M \times K}$ denotes the matrix of coefficients between $\mathbf{x}$ and $\mathbf{y}$. $\sigma_1^2 \mathbf{I}_M$ denotes the variances of the additive noise. $\mathbf{I}_M$ is an identity matrix. Here we drop the intercept terms because the input data $\mathbf{X}$ and $\mathbf{Z}$ are normalized to zero mean and unit variance as preprocessing.

Similarly, the relationship between $\mathbf{y}$ and $\mathbf{z}$ can be represented as

$$\mathbf{z} = \mathbf{By} + \mathbf{Cx} + \boldsymbol{\varepsilon}_2, \tag{7}$$

where

$$\boldsymbol{\varepsilon}_2 \sim \mathcal{N}(\mathbf{0}, \sigma_2^2 \mathbf{I}_N).$$

$\mathbf{B} \in \mathbb{R}^{N \times M}$ denotes the matrix of coefficients between $\mathbf{y}$ and $\mathbf{z}$, $\mathbf{C} \in \mathbb{R}^{N \times K}$ denotes the matrix of coefficients between $\mathbf{x}$ and $\mathbf{z}$ to encode the individual associations.

Note that Eq. 7 decouples the associations between SNPs and genes into two parts: one for individual associations represented as $\mathbf{Cx}$, and another for group-wise associations represented as $\mathbf{By}$. Next, we infer the group-wise associations by a two-layer feature selection strategy. We first remove the individual associations and denote

$$\tilde{\mathbf{Z}} = \mathbf{Z} - \mathbf{CX}. \tag{8}$$

Thus $\tilde{\mathbf{Z}}$ contains only group-wise effects. Next let

$$\mathbf{Y} = \mathbf{AX}. \tag{9}$$

Thus $\mathbf{Y}$ represents a low-rank transformation of the original SNP matrix. Each row of $\mathbf{Y}$ represents a group of SNPs. From Eq. 7, we have the following multiple-input multiple-output (MIMO) linear system

$$\tilde{\mathbf{Z}} = \mathbf{BY} + \mathbf{E}, \tag{10}$$

where $\mathbf{E}$ is a Gaussian white-noise term. In Eqs. 9 and 10, $\mathbf{A}$ and $\mathbf{B}$ should be sparse since a single gene is often influenced by a small number of SNPs and vice versa [16].

Therefore, the overall objective function is

$$\min_{\mathbf{A},\mathbf{B},\mathbf{C},\mathbf{L}} \text{loss}(\mathbf{A},\mathbf{B},\mathbf{C},\mathbf{L}) + \rho||\mathbf{L}||_* + \alpha||\mathbf{A}||_1 + \beta||\mathbf{B}||_1 + \gamma||\mathbf{C}||_1, \tag{11}$$

where α, β, γ, ρ are the regularization parameters, and the loss function is

$$\text{loss}(\mathbf{A},\mathbf{B},\mathbf{C},\mathbf{L}) = ||\mathbf{Z} - \mathbf{L} - (\mathbf{BA} + \mathbf{C})\mathbf{X}||_F^2. \tag{12}$$

Here, we choose different penalties for $\mathbf{A}$, $\mathbf{B}$, $\mathbf{C}$ because the sparsities of different matrices are typically of different scales.

2.3 Optimization

The optimization for $\mathbf{L}$ can be achieved by following a similar approach as in [27]. To optimize $\mathbf{A}$, $\mathbf{B}$, $\mathbf{C}$, many tools can be used to optimize the ℓ_1 penalized objective function, e.g., the Orthant-Wise Limited-memory Quasi-Newton (OWL-QN) algorithm [1]. Due to space limitation, we omit the details. In the next, we devise optimization techniques that can dramatically improve the computational efficiency of geQTL.

2.4 Boosting the Computational Efficiency

Given a large number of SNPs and gene expression traits, scalability of the algorithm is a crucial issue. We propose two improved models, geQTL$^+$ and geQTL-ridge, which optimize the search for significant individual associations, which is the main computational bottleneck of the algorithm.

2.4.1 geQTL$^+$

In a typical eQTL study, we usually have $M \ll \min(K, N)$. Thus, the bottleneck of the algorithm is to optimize $\mathbf{C}$. Our strategy is to confine the space of $\mathbf{C}$. The intuition is that we only permit a small fraction of elements in $\mathbf{C}$ to be nonzero. It has been shown that if $\mathbf{Z}$ and $\mathbf{X}$ are standardized with zero mean and unit sum of squares, then $\mathbf{r} = \text{abs}(\mathbf{ZX}^{\text{T}})$ is equal to the gene-SNP correlations ($\mathbf{r}_{gs} = |\text{cor}(z_g, x_s)|$) [22]. Since for many test statistics, e.g., t, F, R^2, and LR, for the simple linear regression problem can be expressed as functions of the sample correlation $\mathbf{r}_{gs}$, e.g., $R^2 = r^2$, and $t = \frac{r\sqrt{n-2}}{1-r^2}$, we can find a threshold according to the required p-value, such that test statistics exceeding the threshold are significant at the required significance level. The test statistics for every gene-SNP pair in $\mathbf{r}$ are compared with the threshold, and only those elements whose $\mathbf{r}$ are greater than the threshold are optimized. We denote $\mathbf{R}\in\mathbb{R}^{N\times K}$ as the indicator matrix indicating which elements in $\mathbf{C}$ can be nonzero (i.e., $\mathbf{r}_{gs} >$ threshold).

2.4.2 geQTL-Ridge

When N and K are extremely large, optimizing $\mathbf{C}$ may still be time-consuming, since it may take many iterations to converge with the ℓ_1 constraint. Next, we introduce geQTL-ridge, which further improves the time efficiency with slight decrease in accuracy. The key idea is to use ridge regression for individual associations so that we can get a closed form solution for $\mathbf{C}$. The objective function is shown in the following.

$$\begin{aligned} \min_{\mathbf{A},\mathbf{B},\mathbf{C},\mathbf{L}} \quad & \text{loss}(\mathbf{A},\mathbf{B},\mathbf{C},\mathbf{L}) + \rho||\mathbf{L}||_* + \alpha||\mathbf{A}||_1 + \beta||\mathbf{B}||_1 + \gamma||\mathbf{C}||_2^2, \\ \text{s.t.} \quad & (\mathbf{C})_{i,j} \text{ is nozero only if } (\mathbf{R})_{i,j} \text{ is } 1. \end{aligned} \tag{13}$$

Theorem 1: *The solution of* $\mathbf{C}$ *in Eq.*13 *is*

$$\mathbf{c}_i \leftarrow \mathbf{d}_i\mathbf{X}^{\mathrm{T}}\mathbf{P}_i(\mathbf{P}_i^{\mathrm{T}}\mathbf{X}\mathbf{X}^{\mathrm{T}}\mathbf{P}_i + \gamma\mathbf{I}_K)^{-1}\mathbf{P}_i^{\mathrm{T}}, \tag{14}$$

where

$$\mathbf{c}_i = (\mathbf{C})_{i,:}, \mathbf{d}_i = (\mathbf{D})_{i,:},$$

$$\mathbf{D} = \mathbf{Z} - \mathbf{L} - \mathbf{BAX},$$

and $\mathbf{P}_i$ *is defined as in formula* 19.

The proof of the Theorem 1 is in the following section.

2.5 Proof of Theorem 1

Proof: Recall that any ridge regression problem

$$\min_{\mathbf{a}}||\mathbf{b} - \mathbf{aQ}||_2^2 + ||\mathbf{a}\Gamma||_2^2, \tag{15}$$

where $\mathbf{a}$ is a row vector and $\mathbf{Q}$ has linearly independent rows, has the following solution

$$\mathbf{a} = \mathbf{bQ}^{\mathrm{T}}(\mathbf{QQ}^{\mathrm{T}} + \Gamma\Gamma^{\mathrm{T}})^{-1}. \tag{16}$$

Note that

$$\text{loss}(\mathbf{A},\mathbf{B},\mathbf{C},\mathbf{L}) = ||\mathbf{D} - \mathbf{CX}||_F^2 = \sum_{i=1}^{N} ||\mathbf{d}_i - \mathbf{c}_i\mathbf{X}||_2^2, \tag{17}$$

where $\mathbf{D} = \mathbf{Z} - \mathbf{L} - \mathbf{BAX}$, $\mathbf{c}_i = (\mathbf{C})_{i,:}$ and $\mathbf{d}_i = (\mathbf{D})_{i,:}$.

We have

$$\min_{\mathbf{C}} \quad \text{loss}(\mathbf{A}, \mathbf{B}, \mathbf{C}, \mathbf{L}) = \sum_{i=1}^{N} \min_{\mathbf{c}_i} ||\mathbf{d}_i - \mathbf{c}_i\mathbf{X}||_2^2, \tag{18}$$

Taking into account that $(\mathbf{c}_i)_j$ can be nonzero only if $(\mathbf{R})_{i,j}$ is 1, we introduce $\mathbf{P}_i$, where $\mathbf{P}_i$ has K rows and $l_i = \sum_{j=1}^{K} (\mathbf{R})_{i,j}$ columns. And

$$(\mathbf{P}_i)_{s,t} = \begin{cases} 1, & \text{if } (\mathbf{R})_{i,s} \text{ isthe } t\text{-th 1 in } (\mathbf{R})_{i,:}; \\ 0, & \text{other wise.} \end{cases} \tag{19}$$

Then $\mathbf{c}_i = \mathbf{c}_i\mathbf{P}_i\mathbf{P}_i^{\mathrm{T}}$, $||\mathbf{d}_i - \mathbf{c}_i\mathbf{X}||_2^2 + \gamma||\mathbf{c}_i||_2^2 = ||\mathbf{d}_i - (\mathbf{c}_i\mathbf{P}_i) \times (\mathbf{P}_i^{\mathrm{T}}\mathbf{X})||_2^2 + \gamma||\mathbf{c}_i\mathbf{P}_i||_2^2$, and

$$\begin{aligned} &\min_{\mathbf{c}_i} ||\mathbf{d}_i - \mathbf{c}_i\mathbf{X}||_2^2 + \gamma||\mathbf{c}_i||_2^2, \\ &\text{s.t. } (\mathbf{c}_i)_j \text{ is nozero only if } (\mathbf{R})_{i,j} \text{ is } 1, \end{aligned} \tag{20}$$

is solved by

$$\mathbf{c}_i = (\mathbf{c}_i\mathbf{P}_i)\mathbf{P}_i^{\mathrm{T}} = \mathbf{d}_i\mathbf{X}^{\mathrm{T}}\mathbf{P}_i(\mathbf{P}_i^{\mathrm{T}}\mathbf{X}\mathbf{X}^{\mathrm{T}}\mathbf{P}_i + \gamma\mathbf{I}_K)^{-1}\mathbf{P}_i^{\mathrm{T}}. \tag{21}$$

Therefore,

$$\begin{aligned} &\min_{\mathbf{C}} \quad \text{loss}(\mathbf{A}, \mathbf{B}, \mathbf{C}, \mathbf{L}) + \gamma||\mathbf{C}||_2^2, \\ &\text{s.t. } (\mathbf{C})_{i,j} \text{ is nozero only if } (\mathbf{R})_{i,j} \text{ is } 1, \end{aligned}$$

is solved by $\mathbf{C} = (\mathbf{c}_1^{\mathrm{T}}, \ldots, \mathbf{c}_N^{\mathrm{T}})^{\mathrm{T}}$, which leads to the update formula given in Eq. 14.

2.6 Determining the Number of Hidden Variables

In Eq. 12, we use **BA** + **C** to formulate the overall associations between SNPs and expression traits. Two group-wise associations will not share the same group of SNPs (or genes), since otherwise these two group-wise associations can be combined into one. Therefore, every group-wise association should be unique and irreplaceable. Hence, following two conditions should be satisfied

- **A** has linearly independent rows. Since $M \ll K$, this condition is equivalent to that **A** has full rank;
- **B** has linearly independent columns. Since $M \ll N$, this condition is equivalent to that **B** has full rank.

When these two conditions are met, we have

$$M = \text{rank}(\mathbf{A}) = \text{rank}(\mathbf{B}) = \text{rank}(\mathbf{BA}). \tag{22}$$

The last equality holds because both **A** and **B** have full rank.

We have the following observation. The singular value decomposition (SVD) of **BA** has the form

$$\mathbf{BA} = \mathbf{U}\Sigma\mathbf{V}^{\mathrm{T}},$$

where **U** and **V** are unitary (orthogonal in our case) matrices, and Σ is a rectangular diagonal matrix with non-negative real numbers on the diagonal, which corresponds to singular values of **BA**. Since **U** and **V** are unitary and hence have full rank, we have

$$\begin{aligned}\text{rank}(\mathbf{BA}) &= \text{rank}(\mathbf{U}\Sigma\mathbf{V}^{\mathrm{T}}) = \text{rank}(\Sigma)\\ &= \text{the number of nonzero singular values of } \mathbf{BA}.\end{aligned} \tag{23}$$

We compute **BA** by minimizing Eq. 12, which gives

$$\mathbf{BA} = (\mathbf{Z} - \mathbf{L} - \mathbf{CX})\mathbf{X}^{\mathrm{T}}(\mathbf{XX}^{\mathrm{T}})^{-1}. \tag{24}$$

Combine 22, 23, and 24, we find

$$\begin{aligned}M = &\text{ the number of nonzero singular values of}\\ &(\mathbf{Z} - \mathbf{L} - \mathbf{CX})\mathbf{X}^{\mathrm{T}}(\mathbf{XX}^{\mathrm{T}})^{-1}.\end{aligned} \tag{25}$$

Due to the existence of noise, we should allow small singular values to be considered as zero. Therefore, we can draw a plot with singular values of $(\mathbf{Z} - \mathbf{L} - \mathbf{CX})\mathbf{X}^{\mathrm{T}}(\mathbf{XX}^{\mathrm{T}})^{-1}$ in descending order and set M to be k, if the first k singular values are large and significantly greater than the $(k + 1)$-th singular value.

Based on the discussion above, in order to find optimal M, we can first use Lasso to infer the initial value of **C**. Then, using Eq. 25, we can infer the optimal M at this stage. After that, we can optimize new **C** and calculate new optimal M. We can repeat this procedure until M became stable or reach maximal number of iterations.

3 Experimental Study

In this section, we perform extensive experimental study using both simulated and real eQTL datasets to evaluate the performance of our methods. For comparison, we select several state-of-the-art eQTL methods, including two-graph guided multi-task Lasso (MTLasso2G) [6], FaST-LMM [16], SET-eQTL [7], LORS [27], Matrix eQTL [22], and Lasso [25]. Note that we did not compare with our previous work, GDL, in [8] because it needs to incorporate many prior knowledge, that is not relevant to this work. For all the methods, the tuning parameters are learned using cross validation. The discussion of setting proper number of group-wise associations M is included in the supplementary material. The shrinkage of the coefficients is also presented in the supplementary material.

3.1 Simulated Data

We use a similar setup for simulation study to that in [27]. First, 100 SNPs are randomly selected from the yeast eQTL dataset [4]. This gives birth to the matrix **X**. 100 gene expression profiles

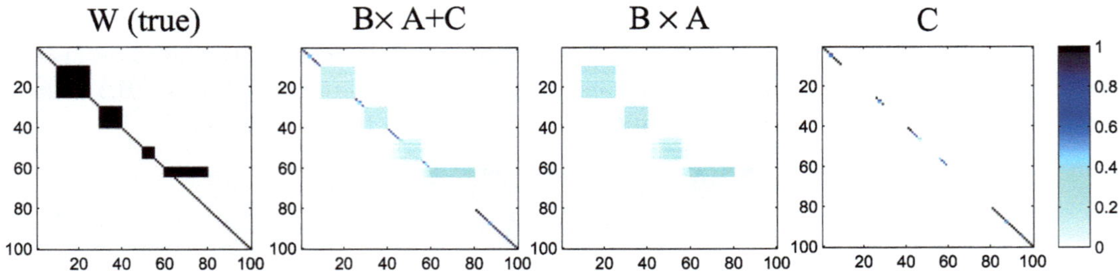

Fig. 2 Ground truth of matrix **W** and associations estimated by geQTL. The *x*-axis represents SNPs and *y*-axis represents traits. Normalized absolute values of regression coefficients are used. Darker color implies stronger association. Number of group-wise associations $M = 4$

are generated by $\mathbf{Z}_{j*} = \boldsymbol{\beta}_{j*}\mathbf{X} + \Xi_{j*} + \mathbf{E}_{j*}$ $(1 \leq j \leq N)$, where $\mathbf{E}_{j*} \sim \mathcal{N}(0, \phi I)$ $(\phi = 0.1)$ is used to simulate the Gaussian noise. To simulate the effects of confounding factors, we use Ξ_{j*}, drawn from $\mathcal{N}(\mathbf{0}, \tau\Lambda)$. In this chapter, we set $\tau = 0.1$. Λ is given by $\mathbf{F}\mathbf{F}^{\mathrm{T}}$. Here, $\mathbf{F} \in \mathbb{R}^{H \times J}$ and $\mathbf{F}_{ij} \sim \mathcal{N}(0, 1)$. J denotes hidden factor number. In this chapter, we set J to 10.

In the left most of Fig. 2, we illustrate $\boldsymbol{\beta}$. Here, we set the association strength to 1. Totally, there exist four group-wise associations with different scales. The diagonal line represents the individual signals in *cis*-regulation.

In Fig. 2, we report the associations inferred by geQTL. Recall that group-wise associations can be inferred from matrix **A** and **B**, and individual associations can be inferred from matrix **C**. It is obvious that geQTL can detect both group-wise and individual signals.

We use $SNR = \sqrt{\frac{Var(\boldsymbol{\beta}\mathbf{X})}{Var(\Xi + \mathbf{E})}}$ to denote the signal-to-noise ratio [27] in the eQTL datasets. Here, we fix $J = 10, \tau = 0.1$. The *SNR*s are controlled by using different ϕs. Using 50 simulated datasets with different *SNR*s, we compare the proposed methods with the selected methods. Because FaST-LMM requires the input of genomic locations information (e.g., chromosome, base pair, etc.), we will compare it on the real dataset. The results are averaged over 50 different simulated datasets. **BA** + **C** is used to represent the association matrix in our method. Figure 3 shows the ROC curve of TPR–FPR (true positive rate–false positive rate) for performance comparison. Typically, we care more about the TPR when the FPR is small because it is important to evaluate the performance of model when controlling the maximum tolerated FPR. Thus, in Fig. 3, the ROC of interest for eQTL is generally shown in the range [0,0.1]. The corresponding areas under the TPR-FPR curve are shown in Fig. 4.

It can be seen that geQTL and geQTL$^+$ outperform all alternative methods by a large margin since they consider both individual and group-wise associations. We also observe that geQTL-ridge is not as good as geQTL and geQTL$^+$. This is because geQTL-ridge

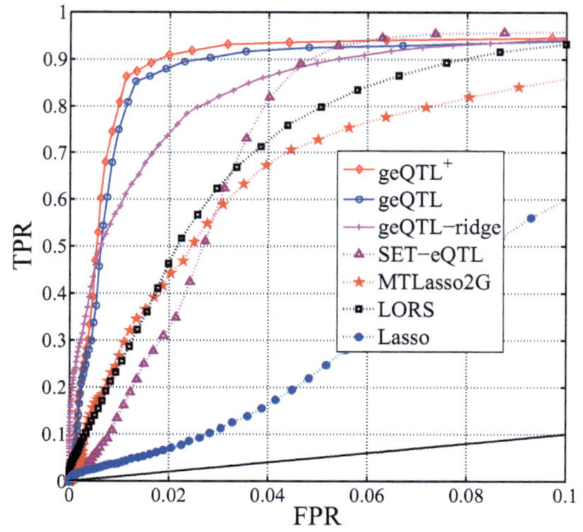

Fig. 3 The ROC curve of FPR-TPR with different signal-to-noise ratios ($SNR = 0.13$)

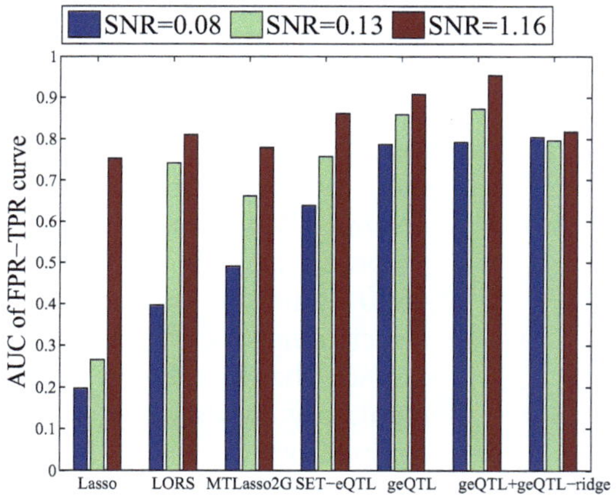

Fig. 4 The AUCs curve

does not provide a sparse solution for individual associations. MTLasso2G is comparable to LORS. LORS can correct the effects of the confounders, however, it is not able to detect group-wise mappings. We also observe that by decoupling individual and group-wise associations, the proposed models (geQTL, geQTL$^+$, and geQTL-ridge) are more robust to noise than other methods.

3.2 Yeast eQTL Data

We also validated *geQTL* using the bench mark dataset–yeast (Saccharomyces cerevisiae) eQTL dataset. The dataset contains 112 yeast segregants generated from a cross of two inbred strains [4]. Originally, It contains 6229 gene expressions and 2956 SNPs. SNPs with >10% missing values in the remaining SNPs are imputed using the function fill.geno in R/QTL [5]. The neighboring SNPs with the same genotype profiles are combined, resulting in 1027 genotype profiles. Remove gene expression traits with missing values, we get 4474 expression profiles.

3.2.1 *cis- and trans-Analysis*

We follow the standard *cis*-enrichment analysis that is used in [15, 18] for evaluation. Moreover, we use the *trans*-enrichment with a similar strategy [28]. Genes regulated by transcription factors (obtained from http://www.yeastract.com/download.php) are treated as *trans*-acting signals.

In Table 2, we report the pairwise comparison using cis- and trans- enrichment analysis. We do not list geQTL separately from geQTL+ , since geQTL+ is a faster version of geQTL. In this table, the methods are sorted (from top to bottom in the left column and from left to right in the top row) in decreasing order of performance. A p-value shows how significant a method on the left column outperforms a method in the top row in terms of *cis*- and *trans*-enrichments. We observe that geQTL^{+} has significantly better *cis*-enrichment scores than the other models. For *trans*-enrichment, geQTL^{+} is the best, and MTLasso2G comes in second, outperforming FaST-LMM, SET-eQTL, LORS, Matrix eQTL, and Lasso. LORS outperforms Matrix eQTL and Lasso for both *cis*- and *trans*-enrichment. This is because LORS considers confounding factors while Matrix eQTL and Lasso do not. In total, these methods each detected about 6000 associations according to nonzero **W** values. We estimate FDR using 50 permutations as proposed in [27]. With FDR $\leq$ 0.01, geQTL^{+} obtains about 4500 significant associations. The plots of all identified significant associations for different methods are given in Fig. 5. Obviously, we can see that $\mathbf{C} + \mathbf{B} \times \mathbf{A}$ and $\mathbf{C}$ of geQTL^{+} report weaker *trans*-regulatory bands while stronger *cis*-regulatory signals than other competitors.

3.2.2 *Gene Ontology Enrichment Analysis on Detected Group-Wise Associations for Yeast*

We further evaluate the quality of detected groups of genes by measuring the correlations between the detected groups of genes and the GO (Gene Ontology) categories [24]. Specifically, the GO enrichment test is calculated by DAVID [12]. In this chapter, gene sets reported by our algorithm with calibrated p-values less than 0.01 are considered as significantly enriched.

Since SET-eQTL is the only previous approach capable of detecting group-wise association mapping, we compare the groups of genes detected by geQTL and those by SET-eQTL. For SET-eQTL, 90 out of 150 gene sets are significantly enriched. By contrast, 28 out of 30 gene sets reported by geQTL are significantly enriched. This well illustrates the effectiveness of geQTL to infer group-wise associations. The number of SNPs in each group reported by geQTL and their genomic locations are shown in Fig. 6. We can clearly observe that SNPs in the same group are often physically close to each other. This is reasonable because

Table 2
Pairwise comparison of different models using *cis*-enrichment and *trans*-enrichment

	FaST-LMM	geQTL-ridge	SET-eQTL	MTLasso2G	LORS	Matrix eQTL	Lasso
cis geQTL$^+$	< 0.0163	0.0124	< 0.0001	< 0.0001	< 0.0001	< 0.0001	< 0.0001
FaST-LMM	–	0.0247	< 0.0001	< 0.0001	< 0.0001	< 0.0001	< 0.0001
geQTL-ridge	–	–	< 0.0001	< 0.0001	< 0.0001	< 0.0001	< 0.0001
SET-eQTL	–	–	–	0.0117	< 0.0001	< 0.0001	< 0.0001
MTLasso2G	–	–	–	–	< 0.0001	< 0.0001	< 0.0001
LORS	–	–	–	–	–	< 0.0001	0.0052
Matrix eQTL	–	–	–	–	–	–	0.0134

	MTLasso2G	FaST-LMM	LORS	SET-eQTL	Matrix eQTL	Lasso	geQTL-ridge
trans geQTL$^+$	0.0042	0.0040	0.0033	0.0029	0.0027	0.0022	0.0001
MTLasso2G	–	0.0212	0.0134	0.0049	0.0042	0.0038	0.0005
FaST-LMM	–	–	0.0233	0.0178	0.0125	0.0073	0.0006
LORS	–	–	–	0.3110	0.1103	0.0151	0.0008
SET-eQTL	–	–	–	–	0.1223	0.0578	0.0016
Matrix eQTL	–	–	–	–	–	0.0672	0.0021
Lasso	–	–	–	–	–	–	0.0025

SNPs nearby usually jointly affect the expression level of a set of genes that achieves a specific cell function [20].

3.2.3 Reproducibility of eQTLs Between Studies

To further evaluate the identified associations, we investigate the consistency of calls between two independent studies [23]. We examine the reproducibility based on the following two criteria [10, 13, 27]:

- Reproducibility of detected SNP-gene associations: Let L_1 and L_2 be the sets of SNP-gene associations detected in the two yeast datasets, respectively. We can rank the associations according to the weights (or q-values for FaST-LMM). Let L_1^T and L_2^T be the top T most significant associations from the two datasets. The reproducibility is defined as $\frac{|L_1^T \bigcap L_2^T|}{T}$.
- Reproducibility of *trans* regulatory hotspots: For each SNP, we count the number of associated genes from the detected

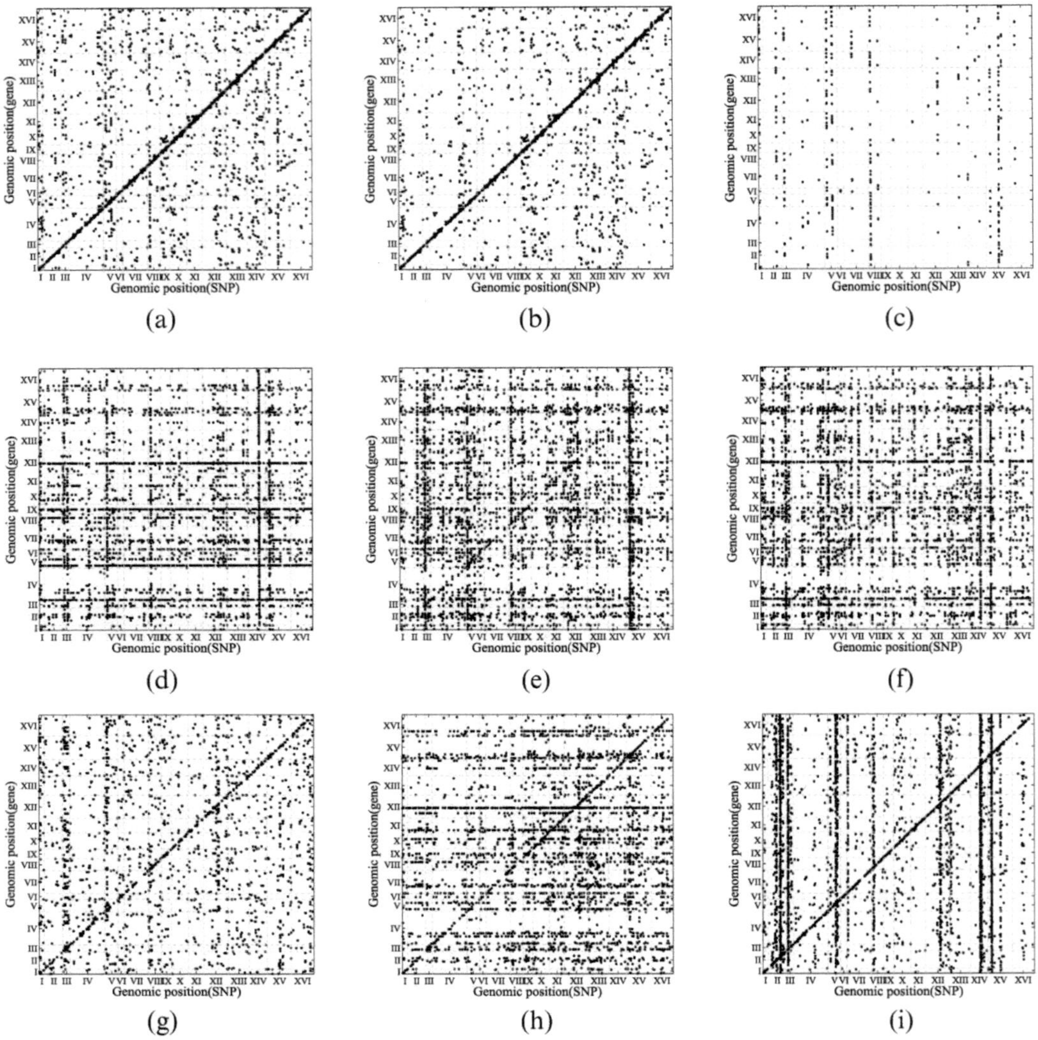

Fig. 5 Significant associations reported on yeast eQTL dataset. (**a**) geQTL$^+$ **C** + **B** $\times$**A** ($M = 30$, top 4500). (**b**) geQTL$^+$ **C** ($M = 30$, top 4000). (**c**) geQTL$^+$ **B** $\times$ **A** ($M = 30$, top 500). (**d**) SET-eQTL ($M = 120$, top 4500). (**e**) SET-eQTL ($M = 150$, top 4500). (**f**) SET-eQTL ($M = 200$, top 4500). (**g**) MTLasso2G (top 4500). (**h**) LORS (top 4500). (**i**) Lasso (top 4500)

SNP-gene associations. We use this number as the *regulatory degree* of each SNP. For FaST-LMM, SNP-Gene pairs with a q-value < 0.001 are considered significant associations. We also tried different cutoffs for FaST-LMM (from 0.01 to 0.001), the results are similar. SNPs with large regulatory degrees are often referred to as hotspots. We sort SNPs in descending order of their regulatory degrees. We denote the sorted SNPs lists as S_1 and S_2 for the two yeast datasets. Let S_1^T and S_2^T be the top T SNPs in the sorted SNP lists. The trans calling consistency of reported hotspots is denoted by $\frac{|S_1^T \bigcap S_2^T|}{T}$.

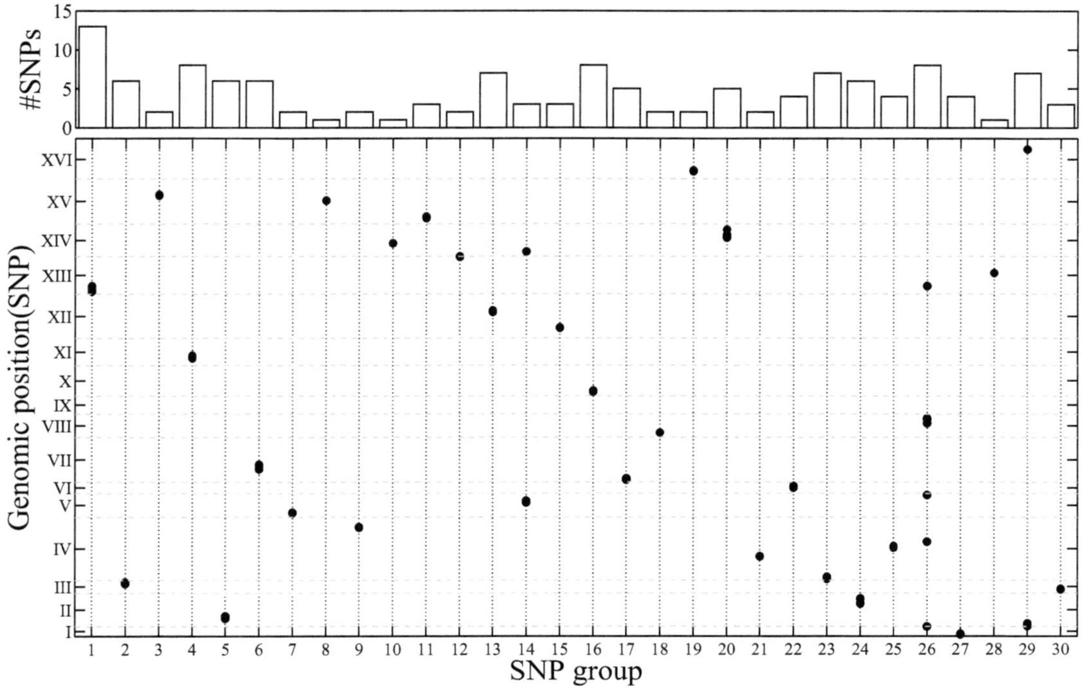

Fig. 6 Genomic positions of SNPs in each SNP group

In Fig. 7 (a), we show the consistency of the top 4500 associations between different studies. In Fig. 7b, we list the reproducibility of *trans* regulatory hotspots reported by different approaches. Overall, geQTL^{+} yielded results with greater consistency all other methods. This well illustrates the superiority of geQTL^{+}.

4 Conclusion

In literature, much efforts have been done on eQTL mapping. Traditional eQTL mapping approaches cannot detect the group-wise associations between sets of SNPs and sets of genes. To achieve that, we propose a fast approach, *geQTL*, to detect both individual and group-wise associations for eQTL mapping. *geQTL* can also correct the effects of potential confounders. We also introduce efficient algorithms to scale up the computation so that the algorithms are able to tackle large scale datasets. Additionally, the proposed model provides an effective strategy to automatically infer the proper number of group-wise associations. We perform extensive experiments on both simulated datasets and yeast datasets to demonstrate the effectiveness and efficiency of the proposed method. Inferring individual and group-wise associations also helps us better explain the genetic basis of gene expression. Due to scalability issue, our model simply assume random errors

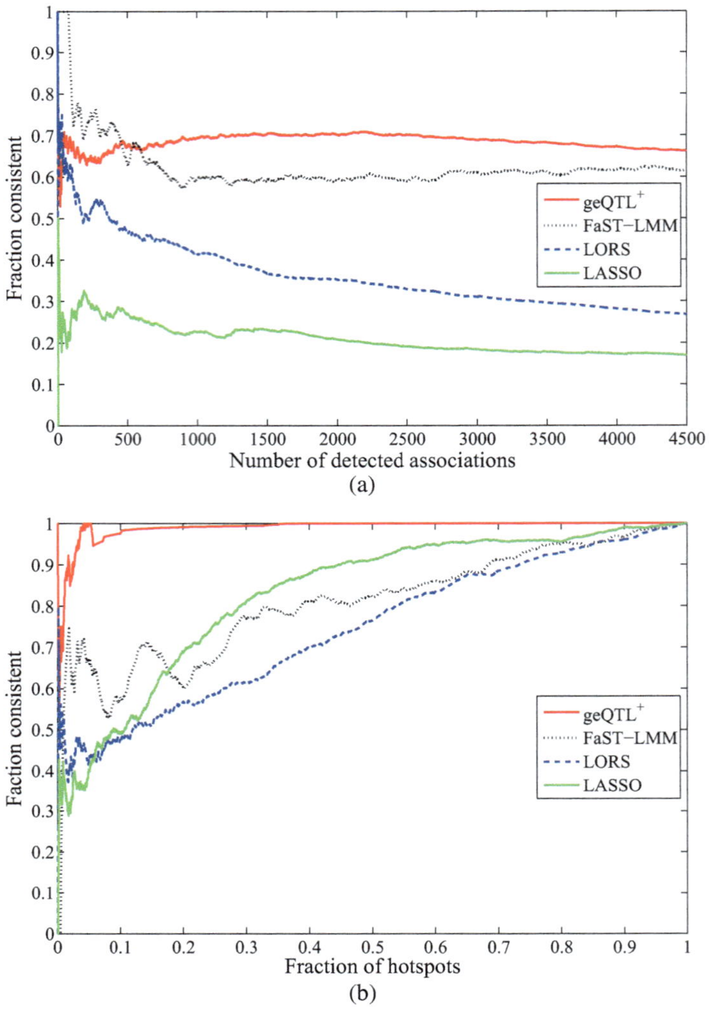

Fig. 7 Reproducibility of eQTLs between two independent yeast eQTL datasets. (**a**) Reproducibility of detected SNP-gene associations; (**b**) reproducibility of *trans* regulatory hotspots reported by different approaches

between different genes are independent and have the same variance. That is the reason why current model only identified a small number of group-wise associations. Our future work will further incorporate the relationships between genes by integrating gene co-expression network or protein–protein interaction network.

References

1. Andrew G, Gao J (2007) Scalable training of L1-regularized log-linear models. In: Proceedings of the 24th international conference on machine learning
2. Bochner BR (2003) New technologies to assess genotype-phenotype relationships. Nat Rev Genet 4:309–314
3. Braun R, Buetow K (2011) Pathways of distinction analysis: a new technique for multi-SNP analysis of GWAS data. PLoS Genet 7 (6):e1002101
4. Brem RB, Storey JD, Whittle J, Kruglyak L (2005) Genetic interactions between

polymorphisms that affect gene expression in yeast. Nature 436:701–703

5. Broman KW, Wu H, Sen S, Churchill GA (2003) R/QTL: QTL mapping in experimental crosses. Bioinformatics 19(7):889–890
6. Chen X, Shi X, Xu X, Wang Z, Mills R, Lee C, Xu J (2012) A two-graph guided multi-task Lasso approach for eQTL mapping. In: 15th International conference on artificial intelligence and statistics, AISTATS 2012, pp 208–217
7. Cheng W, Zhang X, Wu Y, Yin X, Li J, Heckerman D, Wang W (2012) Inferring novel associations between SNP sets and gene sets in eQTL study using sparse graphical model. In: ACM conference on bioinformatics, computational biology and biomedicine '12, pp 466–473
8. Cheng W, Zhang X, Guo Z, Shi Y, Wang W (2014) Graph-regularized dual Lasso for robust eQTL mapping. Bioinformatics 30 (12):i139–i148
9. Cheung VG, Spielman RS, Ewens KG, Weber TM, Morley M, Burdick JT (2005) Mapping determinants of human gene expression by regional and genome-wide association. Nature 437:1365–1369
10. Fusi N, Stegle O, Lawrence ND (2012) Joint modelling of confounding factors and prominent genetic regulators provides increased accuracy in genetical genomics studies. PLoS Comput Biol 8(1):e1002330
11. Holden M, Deng S, Wojnowski L, Kulle B (2008) GSEA-SNP: applying gene set enrichment analysis to SNP data from genome-wide association studies. Bioinformatics 24 (23):2784–2785
12. Huang da W, Sherman BT, Lempicki RA (2009) Systematic and integrative analysis of large gene lists using DAVID bioinformatics resources. Nat Protoc 4(1):44–57
13. Joo JW, Sul JH, Han B, Ye C, Eskin E (2014) Effectively identifying regulatory hotspots while capturing expression heterogeneity in gene expression studies. Genome Biol 15(4): r61
14. Lander ES (2011) Initial impact of the sequencing of the human genome. Nature 470(7333):187–197
15. Listgarten J, Kadie C, Schadt EE, Heckerman D (2010) Correction for hidden confounders in the genetic analysis of gene expression. Proc Natl Acad Sci USA 107(38):16465–16470
16. Listgarten J, Lippert C, Kang EY, Xiang J, Kadie CM, Heckerman D (2013) A powerful and efficient set test for genetic markers that handles confounders. Bioinformatics 29 (12):1526–1533
17. Mazumder R, Hastie T, Tibshirani R (2010) Spectral regularization algorithms for learning large incomplete matrices. J Mach Learn Res 11:2287–2322
18. McClurg P, Janes J, Wu C, Delano DL, Walker JR, Batalov S, Takahashi JS, Shimomura K, Kohsaka A, Bass J, Wiltshire T, Su AI (2007) Genomewide association analysis in diverse inbred mice: power and population structure. Genetics 176(1):675–683
19. Michaelson J, Loguercio S, Beyer A (2009) Detection and interpretation of expression quantitative trait loci (eQTL). Methods 48 (3):265–276
20. Musani SK, Shriner D, Liu N, Feng R, Coffey CS, Yi N, Tiwari HK, Allison DB (2007) Detection of gene x gene interactions in genome-wide association studies of human population data. Hum Hered 63(2):67–84
21. Pujana MA, Han J-DJ, Starita LM, Stevens KN, Tewari M et al. (2007) Network modeling links breast cancer susceptibility and centrosome dysfunction. Nat Genet 39 (11):1338–1349
22. Shabalin AA (2012) Matrix eQTL: ultra fast eQTL analysis via large matrix operations. Bioinformatics 28(10):1353–1358
23. Smith EN, Kruglyak L (2008) Gene-environment interaction in yeast gene expression. PLoS Biol 6(4):e83
24. The Gene Ontology Consortium (2000) Gene ontology: tool for the unification of biology. Nat Genet 25(1):25–29
25. Tibshirani R (1996) Regression shrinkage and selection via the Lasso. J R Stat Soc B 58 (1):267–288
26. Wu MC, Lee S, Cai T, Li Y, Boehnke M, Lin X (2011) Rare-variant association testing for sequencing data with the sequence kernel association test. Am J Hum Genet 89(1):82–93
27. Yang C, Wang L, Zhang S, Zhao H (2013) Accounting for non-genetic factors by low-rank representation and sparse regression for eQTL mapping. Bioinformatics 29 (8):1026–1034
28. Yvert G, Brem RB, Whittle J, Akey JM, Foss E, Smith EN, Mackelprang R, Kruglyak L (2003) Trans-acting regulatory variation in Saccharomyces cerevisiae and the role of transcription factors. Nat Genet 35(1):57–64
29. Zhu J, Zhang B, Smith EN, Drees B, Brem RB, Kruglyak L, Bumgarner RE, Schadt EE (2008) Integrating large-scale functional genomic data to dissect the complexity of yeast regulatory networks. Nat Genet 40(7):854–861

Chapter 9

Exploring Bayesian Approaches to eQTL Mapping Through Probabilistic Programming

Dimitrios V. Vavoulis

Abstract

The discovery of genomic polymorphisms influencing gene expression (also known as *expression quantitative trait loci* or *eQTLs*) can be formulated as a sparse Bayesian multivariate/multiple regression problem. An important aspect in the development of such models is the implementation of bespoke inference methodologies, a process which can become quite laborious, when multiple candidate models are being considered. We describe automatic, black-box inference in such models using `Stan`, a popular probabilistic programming language. The utilization of systems like `Stan` can facilitate model prototyping and testing, thus accelerating the data modeling process. The code described in this chapter can be found at https://github.com/dvav/eQTLBookChapter.

Key words Bayesian variable selection, Global-local shrinkage, Horseshoe prior, RNA-seq, eQTL mapping, Probabilistic programming, Stan, R, Black-box Bayesian inference

1 Introduction

The study of genomic variation and its association with gene expression is critical for elucidating the genetic basis of complex traits, including diseases. The advent of next-generation sequencing (NGS) made possible the detailed investigation of this relationship (also known as *eQTL mapping*) in large cohorts, but it also gave rise to novel statistical challenges [1, 2]. eQTL mapping can be examined in the context of sparse Bayesian multivariate/multiple regression, where we typically propose a number of candidate statistical models followed by benchmarking them against each other in terms of computational and statistical efficiency [3]. The most laborious aspect of this process is the development and software implementation of statistical inference algorithms for each model under consideration, a task that can be impeded by the potential absence of favorable mathematical properties (e.g., conjugacy) in any of these models.

Xinghua Mindy Shi (ed.), *eQTL Analysis: Methods and Protocols*, Methods in Molecular Biology, vol. 2082,
https://doi.org/10.1007/978-1-0716-0026-9_9,

The aim of this chapter is to demonstrate the utility of a popular probabilistic programming language (PPL), `Stan`, in the prototyping and testing phases of the data modeling process for eQTL mapping [4]. PPLs make it possible to describe and perform inference in hierarchical probabilistic models in a matter of minutes using a small amount of high-level code. Besides `Stan`, another popular PPL is `PyMC`, a Python-based software offering excellent performance, a wide range of inference algorithms, and the ability to mix these algorithms in the same inferential cycle for estimating different parts of a given model [5]. The reason for choosing `Stan` in this chapter is its simplicity, which stems from the fact that its syntax is very close to the mathematical notation already familiar to statisticians.

The practical part of this chapter covers: (a) the acquisition of genomic variation and gene expression data from online sources, (b) the use of these to simulate artificial eQTL datasets with known properties, (c) the implementation of statistical models in `Stan` for eQTL mapping, and (d) the estimation of unknown model parameters using the previously simulated datasets. In the remaining part of this introduction, we outline the statistical theory, which underpins the practical part of this chapter.

1.1 Theory

We assume an $N \times M$ matrix $Z = \{z_{ij}\}$ of read counts quantifying the expression of N transcripts in M samples, an $N \times M$ matrix $C = \{c_{ij}\}$ of transcript- and sample-specific normalization factors, and an $M \times K$ matrix $X = \{x_{jk}\}$ of genotypes indicating the number of minor alleles (0, 1 or 2) in each of K bi-allelic genomic loci in M samples. A matrix $\tilde{X} = \{\tilde{x}_{jk}\}$ is derived by standardizing each column of X.

We introduce an $N \times K$ sparse matrix B of regression coefficients with elements β_{ik}, which measure the effect of variant k on the expression of transcript i. Estimating B is the main focus of subsequent analysis. Typically, sparsity is induced on B by imposing appropriate priors. A common sparsity-inducing prior is a two-component mixture of the following form:

$$\beta_{ik} \sim (1 - \pi_{ik})\delta_0 + \pi_{ik}\mathcal{N}(0, \sigma_i^2)$$

where δ_0 is a point-mass distribution centered at 0. This prior can set β_{ik} exactly at 0, but posterior estimation requires stochastic exploration of a 2^{NK}-dimensional discrete space. An alternative approach with obvious computational advantages would be to adopt a continuous shrinkage prior; for example, the important class of global-local shrinkage priors [6]:

$$\beta_{ik} \sim \mathcal{N}(0, \eta^2\zeta_{ik}^2) \quad \eta^2 \sim \mathrm{p}_\eta(\eta^2) \quad \zeta_{ik}^2 \sim \mathrm{p}_\zeta(\zeta_{ik}^2)$$

where ζ_{ik} are local (i.e., gene- and variant-specific) shrinkage scales, while η is a global scale controlling the overall shrinkage of B. The shrinkage profile of B depends on the form of p_η and p_ζ. Different choices of these distributions give rise to different shrinkage priors, but here we shall use the *horseshoe prior* [7] for which ζ_{ik} and η follow standard and scaled half-Cauchy distributions, $\mathcal{C}^+(0, 1)$ and $\mathcal{C}^+(0, \alpha)$, respectively.

1.1.1 Normal Model

For the Normal model, we assume that read counts have been normalized, $\tilde{z}_{ij} = \frac{z_{ij}}{c_{ij}}$, and log-transformed, $y_{ij} = \log(\tilde{z}_{ij} + 1)$. The model takes the following form:

$$y_{ij} \sim \mathcal{N}\left(\beta_{0i} + \sum_k \beta_{ik}\tilde{x}_{jk}, \sigma_i^2\right) \quad \beta_{0i} \sim 1 \quad \log\sigma_i^2 \sim 1$$

$$\beta_{ik} \sim \mathcal{N}\left(0, \frac{\overline{\sigma}^2}{NK}\eta^2\zeta_{ik}^2\right) \quad \eta \sim \mathcal{C}^+(0, 1) \quad \zeta_{ik} \sim \mathcal{C}^+(0, 1)$$

where σ_i^2 is the variance of gene i and $\sum_k \beta_{ik}\tilde{x}_{jk}$ is the effect of genomic variation on the baseline expression β_{0i} of gene i. We assume that M is large, so we can afford to impose a flat prior distribution on β_{0i} and $\log\sigma_i$ over the interval $(-\infty, +\infty)$. A more complicated model would include correlations between different y_{ij} variables, additional (e.g., clinical, environmental, and population) covariates influencing the baseline gene expression β_{0i}, as well as hierarchical priors on β_{0i} and σ_i^2. Notice that the variance of β_{ik} is proportional to $\overline{\sigma}^2 = \left(\frac{\sum_i \sigma_i}{N}\right)^2$ and inversely proportional to the total number of coefficients. The implicit assumption under this formulation is that the true global scale parameter is $\xi \sim \mathcal{C}^+\left(0, \frac{\overline{\sigma}}{\sqrt{NK}}\right)$. Finally, in this and the subsequent models, we assume that we can ignore any gene-specific factors (e.g., length) affecting c_{ij}, thus $c_{ij} \equiv c_j$.

1.1.2 Negative Binomial Model

The negative binomial distribution is immensely popular for modeling over-dispersed RNA-seq data [8–10], but the mathematical complexities associated with inference in this model might explain (at least partially) the popularity of transformation-based methods, such as `voom` [11]. Here, we examine the following model:

$$z_{ij} \sim \mathcal{NB}(m_{ij}, \phi_i) \quad \log m_{ij} = \log c_{ij} + \log L_j + \beta_{0i} + \sum_k \beta_{ik}\tilde{x}_{jk} \quad \beta_{0i} \sim 1 \quad \log\phi_i \sim 1$$

$$\beta_{ik} \sim \mathcal{N}\left(0, \frac{\eta^2\zeta_{ik}^2}{NK}\right) \quad \eta \sim \mathcal{C}^+(0, 1) \quad \zeta_{ik} \sim \mathcal{C}^+(0, 1)$$

where m_{ij} is the gene- and sample-specific mean of the negative binomial distribution, and ϕ_i is a gene-specific dispersion parameter, such that $\text{Var}[z_{ij}] = m_{ij} + m_{ij}^2\phi_i$; $L_j = \Sigma_i z_{ij}$ is the total number of reads in sample j.

1.1.3 Poisson-LogNormal Model

An alternative approach to work with non-transformed data is to impose a Poisson observational model on top of the Normal:

$$z_{ij} \sim \mathcal{P}(m_{ij}) \quad \log m_{ij} = \log c_{ij} + \log L_j + y_{ij}$$

where y_{ij} serve as latent variables following the Normal model. The Poisson-LogNormal model is motivated by (a) the fact that the Negative Binomial model can be thought of as a Poisson-Gamma mixture and (b) replacing the gamma distribution in the previous mixture with LogNormal. By invoking the law of total expectation and the law of total variance, we can see that $E[z_{ij}] = c_{ij}L_j e^{\beta_{0i}+\sum_k \beta_{ik}\tilde{x}_{jk}+\sigma_i^2/2}$ and $\text{Var}[z_{ij}] = E[z_{ij}] + E[z_{ij}]^2\phi_i$. Hence, the variance has the same form as in the Negative Binomial model, with dispersion parameter $\phi_i = e^{\sigma_i^2} - 1$. When $\sigma_i^2 = 0$, the above model reduces to Poisson.

2 Materials

2.1 Operating System

1. A working UNIX environment (e.g., Linux or MacOS X) with a terminal emulator running the command shell `bash`. The work presented in this chapter was tested on Ubuntu Linux v18.04.

2.2 Software

1. A recent version of the command line tool and library `curl` for transferring data with URLs.
2. A recent version of `vcftools`, a set of tools written in Perl and C++ for working with VCF files [12].
3. A recent version of `R`, the free software environment for statistical computing [13].
4. A recent version of `rstudio`, an integrated development environment for working with `R` and the command line [14].
5. A recent version of `rstan`, an `R` interface for `stan`.
6. A recent version of `plyr`, a set of `R` tools for splitting, modifying, and combining data [15].
7. A recent version of `doMC`, an `R` package providing a parallel backend for multicore computation.
8. A recent version of `cowplot`, an `R` package for plotting.
9. A recent version of `tidyverse`, a collection of `R` packages for data wrangling and plotting.
10. A recent version of `reshape2`, an `R` package for restructuring and aggregating data [16].

The above R packages can be installed either through the graphical interface provided by rstudio, or through the R console; for example install.packages(tidyverse)

3 Methods

3.1 Data Acquisition

1. Start rstudio
2. Create a working directory tree by typing the following commands at the bash command prompt (*see* **Note 1**):

```
mkdir -p eQTLchapter/{data,R,stan}
```

3. Make the root of the tree you just created your working directory by typing the following at the R console (*see* **Note 2**):

```
setwd('eQTLchapter')
```

4. Download genomic variation data from the 1000 Genomes project [17]. At the bash command prompt, type the following (*see* **Note 3**):

```
cd eQTLchapter
curl -o - ftp://ftp.1000genomes.ebi.ac.uk/vol1/ftp/release/20130502/ALL.chr7.phase3_shapeit2_
      mvncall_integrated_v5a.20130502.genotypes.vcf.gz   \
   | vcftools --gzvcf - \
      --chr 7 --from-bp 100000 --to-bp 200000 \
      --remove-filtered-all --remove-indels \
      --min-alleles 2 --max-alleles 2 \
      --maf 0.05 --max-maf 0.95 \
      --012 --out data/chr7
```

5. Download RNA-seq read count data from the ReCount project [18]. At the bash prompt, type the following:

```
curl -o data/montpick_count_table.txt http://bowtie-bio.sourceforge.net/recount/countTables/
montpick_count_table.txt
curl -o data/montpick_phenodata.txt http://bowtie-bio.sourceforge.net/recount/phenotypeTables/
montpick_phenodata.txt
```

3.2 Data Importing

1. Create the following R scripts. At the bash prompt, type (*see* **Note 4**):

```
touch analysis.R R/{utils,viz}.R
```

2. Using the code editor provided by rstudio, add the line library(tidyverse) at the top of utils.R and viz.R, and the following lines at the top of analysis.R (*see* **Note 5**):

```
source('R/utils.R')
source('R/viz.R')

doMC::registerDoMC(cores=8)
```

3. Create a function in `utils.R` for importing the genotypic data into R (*see* **Note 6**):

```
load_genotypes = function() {
  # read sample names
  samples =
    file.path('data', 'chr7.012.indv') %>%
    read_tsv(col_names = 'SAMPLE') %>%
    pull(SAMPLE)

  # read loci
  pos =
    file.path('data', 'chr7.012.pos') %>%
    read_tsv(col_names = c('CHROM', 'POS')) %>%
    unite('POS', CHROM, POS, sep = ':') %>%
    pull(POS)

  # read genotypes
  geno =
    file.path('data', 'chr7.012') %>%
    read_tsv(na = '-1', col_names = c('SAMPLE', pos)) %>%
    select(-1) %>%
    as.matrix()
  dimnames(geno) = list(samples = samples, variants = pos)

  # remove loci with the same genotype across all samples
  sds = apply(geno, 2, sd, na.rm = T)
  geno[,sds > 0]
}
```

4. Source `utils.R`, add the following line to `analysis.R` and execute it in order to import the genotypic data into R:

```
obs_geno = load_genotypes()
```

5. Create the following function in `viz.R` for visualizing the correlation structure of the matrix of genotypes:

```
plot_genotypes = function(geno) {
  geno %>%
    cor() %>%
    reshape2::melt() %>%
    ggplot() +
    geom_raster(aes(x = Var1, y = Var2, fill = value)) +
    theme(legend.position = 'none',
          axis.text.x = element_blank(),
          axis.text.y = element_blank(),
          axis.ticks.x = element_blank(),
          axis.ticks.y = element_blank()) +
    scale_fill_gradient(low = 'black', high = 'white') +
    labs(x = 'genomic position', y = 'genomic position')
}
```

6. Source viz.R, add the following line to analysis.R and execute it in order to visualize the genotypic data (Fig. 1):

```
plot_genotypes(obs_geno)
```

7. Create a function in utils.R for importing the read counts data into R (*see* **Note 7**):

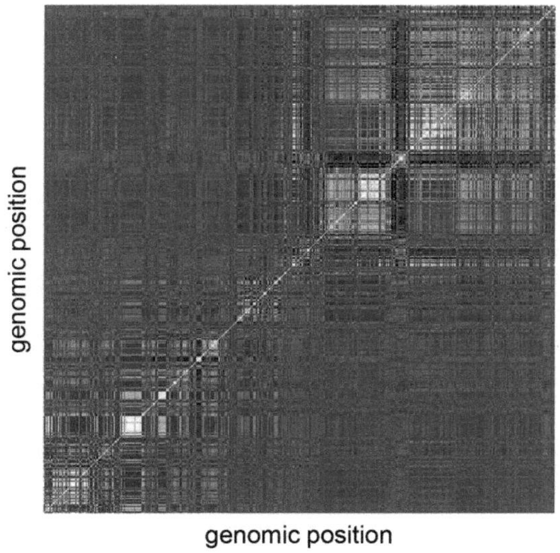

Fig. 1 Correlation structure of the matrix of genotypes. Blocks of highly correlated variants appear in white

```
load_counts = function(pop = 'CEU') {
  # load samples
  samples =
    read_delim(file.path('data', 'montpick_phenodata.txt'), delim = ' ') %>%
    filter(population == pop) %>%
    pull(sample.id)

  # load read counts
  counts =
    read_tsv(file.path('data', 'montpick_count_table.txt')) %>%
    select(GENE = gene, samples) %>%
    as.data.frame() %>%
    column_to_rownames('GENE') %>%
    as.matrix()
  dimnames(counts) = list(genes = rownames(counts), samples = samples)

  # remove genes with the same number of counts across all samples
  sds = apply(counts, 1, sd, na.rm = T)
  counts[sds > 0,]
}
```

8. Source `utils.R`, add the following line to `analysis.R` and execute it in order to import the read counts data into `R`:

```
obs_counts = load_counts()
```

3.3 Estimation of Gene Expression Statistics

1. Create a function in `utils.R` for calculating sample-specific normalization factors given a matrix of count data (*see* **Note 8**):

```
calculate_norm_factors = function(counts) {
  lcounts = log(counts)
  lgm = rowSums(lcounts) / ncol(lcounts)
  idxs = is.finite(lgm)
  lratios = sweep(lcounts[idxs,], 1, lgm[idxs], '-')
  apply(exp(lratios), 2, median)
}
```

2. Create a function in `utils.R` for calculating the log-likelihood of a vector of counts, assuming each element is sampled from the negative binomial distribution (*see* **Note 9**):

```
lognbinom = function(pars, z, cc) {
  mu = pars[1]
  phi = pars[2]

  m = cc * mu
  alpha = 1 / phi
  ll = lgamma(z + alpha) - lgamma(alpha) - lgamma(z + 1) +
    z * log(m) + alpha * log(alpha) - (z + alpha) * log(m + alpha)

  sum(ll)
}
```

3. Create a function in `utils.R` for calculating gene-specific mean and dispersion maximum likelihood estimates for the negative binomial distribution given a matrix of count data (*see* **Note 10**):

```
calculate_gene_stats = function(counts) {
  sizes = calculate_norm_factors(counts)
  optim = possibly(optim, otherwise = NULL)
  stats =
    counts %>%
    plyr::alply(1, function(cnts) {
      optim(par = c(1, 1),
            fn = lognbinom,
            z = cnts,
            cc = sizes,
            control = list(fnscale = -1),
            method = 'L-BFGS-B',
            hessian = T,
            lower = 1e-12)
    }, .parallel = T, .dims = T) %>%
    compact() %>%
    discard(~.x$convergence > 0) %>%
    map('par') %>%
    enframe() %>%
    mutate(value = map(value, str_c, collapse = ',')) %>%
    separate(value, into = c('MEAN', 'PHI'), sep = ',', convert = T) %>%
    mutate(VAR = MEAN + PHI * MEAN^2) %>%
    rename(genes = name) %>%
    as.data.frame() %>%
    column_to_rownames('genes') %>%
    as.matrix()
  dimnames(stats) = list(genes = rownames(stats), statistics = colnames(stats))
  stats
}
```

4. Source `utils.R`, add the following line to `analysis.R` and execute it in order to calculate basic gene-wise statistics from the read counts data:

```
obs_counts_stats = calculate_gene_stats(obs_counts)
```

5. Create a function in `viz.R` for visualizing empirical and estimated gene-wise statistics (*see* **Note 11**):

```
plot_mean_variance = function(counts, stats) {
  plot_fn = function(df, xlabel, ylabel) {
    df %>%
      ggplot() +
      geom_point(aes(x = MEAN, y = VAR), size = 0.1) +
      geom_smooth(aes(x = MEAN, y = VAR), size = 0.5,
                  color = 'red', method = 'lm', formula = y ~ poly(x, 2)) +
      scale_x_continuous(trans = 'log10') +
      scale_y_continuous(trans = 'log10') +
      labs(x = xlabel, y = ylabel)
  }

  # normalise count data
  sizes = calculate_norm_factors(counts)
  counts = sweep(counts, 2, sizes, '/')

  # plot of estimated stats
  gg1 =
    stats %>%
    as.data.frame() %>%
    plot_fn(xlabel = 'estimated mean', ylabel = 'estimated variance')

  # plot of observed stats
  gg2 =
    data.frame(
      MEAN = rowMeans(counts),
      VAR = apply(counts, 1, var)
    ) %>%
    plot_fn(xlabel = 'observed mean', ylabel = 'observed variance')

  # combine plots
  cowplot::plot_grid(gg1, gg2, align = 'vh', labels = 'AUTO')
}
```

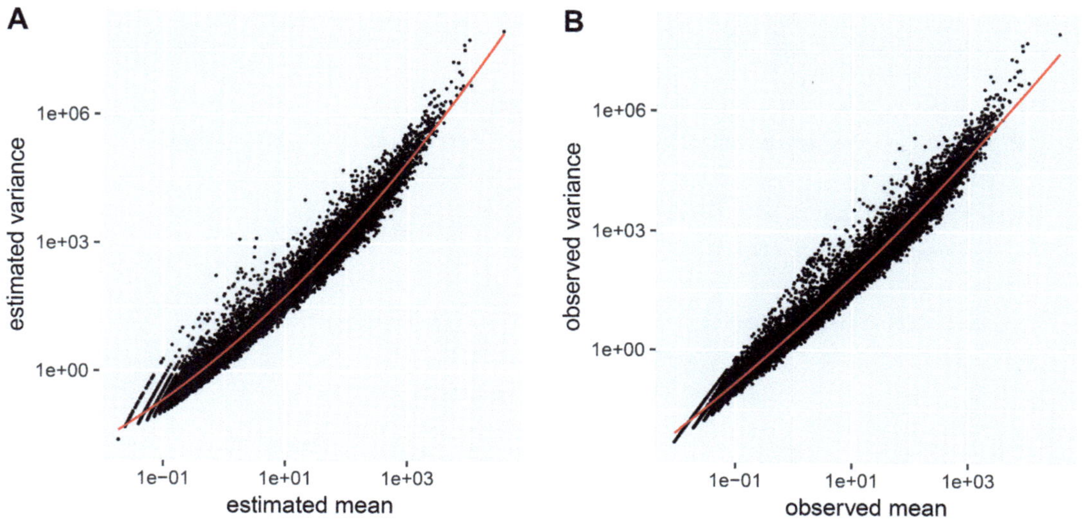

Fig. 2 (**A**) Estimated and (**B**) observed mean-variance relationship in the gene expression data

6. Source `viz.R`, add the following line to `analysis.R` and execute it in order to visualize estimated and observed gene-wise statistics (Fig. 2):

```
plot_mean_variance(obs_counts, obs_counts_stats)
```

3.4 Data Simulation

1. Create a function in `utils.R` for simulating genotypic and count gene expression data given observed data of the same kind and their estimated statistics (*see* **Note 12**):

```
simulate_data = function(count_stats, genotypes, nsamples = 1000, ngenes = 100,
                         nvars = 50, nhits = 10, rate = 4) {
  # fetch genotypes
  X = genotypes[sample(1:nrow(genotypes), nsamples),
                sample(1:ncol(genotypes), nvars)]
  X = X[,apply(X, 2, sd, na.rm = T) > 0]
  dimnames(X) = list(samples = str_c('S', 1:nrow(X)),
                     variants = str_c('V', 1:ncol(X)))

  # simulate matrix of coefficients
  B = matrix(0, nrow = ngenes, ncol = ncol(X),
             dimnames = list(genes = str_c('G', 1:ngenes),
                             variants = str_c('V', 1:ncol(X))))
  hits = c(1 + rexp(0.5 * nhits, rate = rate),
          -1 - rexp(0.5 * nhits, rate = rate))
  B[sample(length(B), length(hits))] = hits
```

```

  # simulate counts
  # df = count_stats[sample(1:nrow(count_stats), ngenes),]
  df = count_stats[order(count_stats[,'MEAN'], decreasing = T),][1:ngenes,]

  X0 = scale(X, center = T, scale = T)
  m = sweep(exp(B %*% t(X0)), 1, df[,'MEAN'], '*')
  alpha = 1 / df[,'PHI']
  Z = matrix(rnbinom(length(m), mu = m, size = alpha), nrow = nrow(m))
  dimnames(Z) = list(genes = str_c('G', 1:nrow(Z)),
                     samples = str_c('S', 1:ncol(Z)))

  # output
  lst(B, X, Z, stats = df)
}
```

2. Source `utils.R` to make `simulate_data` visible to the global environment.

3.5 Model Implementation

1. Copy the files implementing the Normal, Poisson, and Negative Binomial models to the `stan/` directory (`normal.stan`, `poissonln.stan` and `negbinom.stan`, respectively; see end of chapter for code listings) (*see* **Note 13**).
2. Compile the models by adding the following lines in `analysis.R` and executing them (*see* **Note 14**):

```
normal =
  file.path('stan', 'normal.stan') %>%
  rstan::stan_model(auto_write = T)
poissonln =
  file.path('stan', 'poissonln.stan') %>%
  rstan::stan_model(auto_write = T)
negbinom =
  file.path('stan', 'negbinom.stan') %>%
  rstan::stan_model(auto_write = T)
```

3.6 Model Testing

1. Create an auxiliary function in `utils.R` for fitting a compiled model and extracting the estimated regression coefficients B (*see* **Note 15**):

```
fit_model = function(data, model, fcts = calculate_norm_factors(data$Z), ...) {
  Z_tilde = sweep(data$Z, 2, fcts, '/')          # normalise count data
  Y = log(Z_tilde + 1)                           # transform normalised count data
  X0 = scale(data$X, center = T, scale = T)      # standardise genotypes

  # MAP estimation
  fit = rstan::optimizing(object = model,
                          data = list(Z = data$Z,
                                      Y = Y,
                                      X = X0,
                                      c = fcts,
                                      s = colSums(data$Z),
                                      N = nrow(Y),
                                      M = ncol(Y),
                                      K = ncol(X0)),
                          seed = 42,
                          ...)

  # extract estimated matrix of regression coefficients B in vector format
  tibble(EST =
           fit %>%
           pluck('par') %>%
            enframe() %>%
            filter(str_detect(name, '^B\\[')) %>%
            pull(value),
         TRU = as.vector(data$B),
         IDX = 1:length(TRU))
}
```

2. Source `utils.R` to make `fit_model` visible to the global environment.
3. Make the function `cpp_object_initializer` visible to the global environment by adding the following line to `analysis.R` and executing it (*see* **Note 16**):

```
cpp_object_initializer = rstan::cpp_object_initializer
```

4. Set the initial seed of `R`'s random number generator by adding the following line in `analysis.R` and executing it (*see* **Note 17**):

```
set.seed(42)
```

5. Test all three models on simulated data of various sizes (312, 625, 1250, 2500) by adding the following lines in `analysis.R` and executing them (*see* **Note 18**):

```
fitted_models =
  plyr::ldply(lst(312, 625, 1250, 2500), function(N) {
    sim = simulate_data(obs_counts_stats, obs_geno, nsamples = N)
    plyr::ldply(lst(normal, poissonln, negbinom), function(mdl) {
      fit_model(sim, mdl)
    }, .parallel = T, .id = 'MODEL')
  }, .progress = 'text', .id = 'NSAMPLES') %>%
  as_tibble()
```

6. Create a function in `viz.R` for visualizing the results (*see* **Note 19**):

```
plot_fitted_models = function(df, thr = 0.1 * max(abs(df$EST))) {
  df %>%
    mutate(TRU = na_if(TRU, 0),
           EST = if_else(abs(EST) < thr, NA_real_, EST)) %>%
    ggplot() +
    geom_ribbon(aes(x = IDX, ymin = -thr, ymax = thr), fill = 'grey85') +
    geom_hline(yintercept = 0, linetype = 'dashed', size = 0.2) +
    geom_point(aes(x = IDX, y = TRU), color = 'red') +
    geom_point(aes(x = IDX, y = EST), size = 0.5) +
    facet_grid(NSAMPLES~MODEL) +
    scale_x_continuous(expand = c(0.02, 0.02)) +
    scale_y_continuous(expand = c(0.02, 0.02)) +
    labs(x = 'index of regression coefficients (genes x variants)',
         y = 'value of regression coefficients')
}
```

7. Source `viz.R` to make `plot_fitted_models.R` visible to the global environment.

8. Add the following line in `analysis.R` and execute it to visualize the estimated coefficients of the fitted models (Fig. 3):

```
plot_fitted_models(fitted_models)
```

4 Conclusion

Although it is not prudent to make general statements based on a single set of simulations, we may observe from Fig. 3 that all three models perform similarly. At small samples ($N = 312$), the

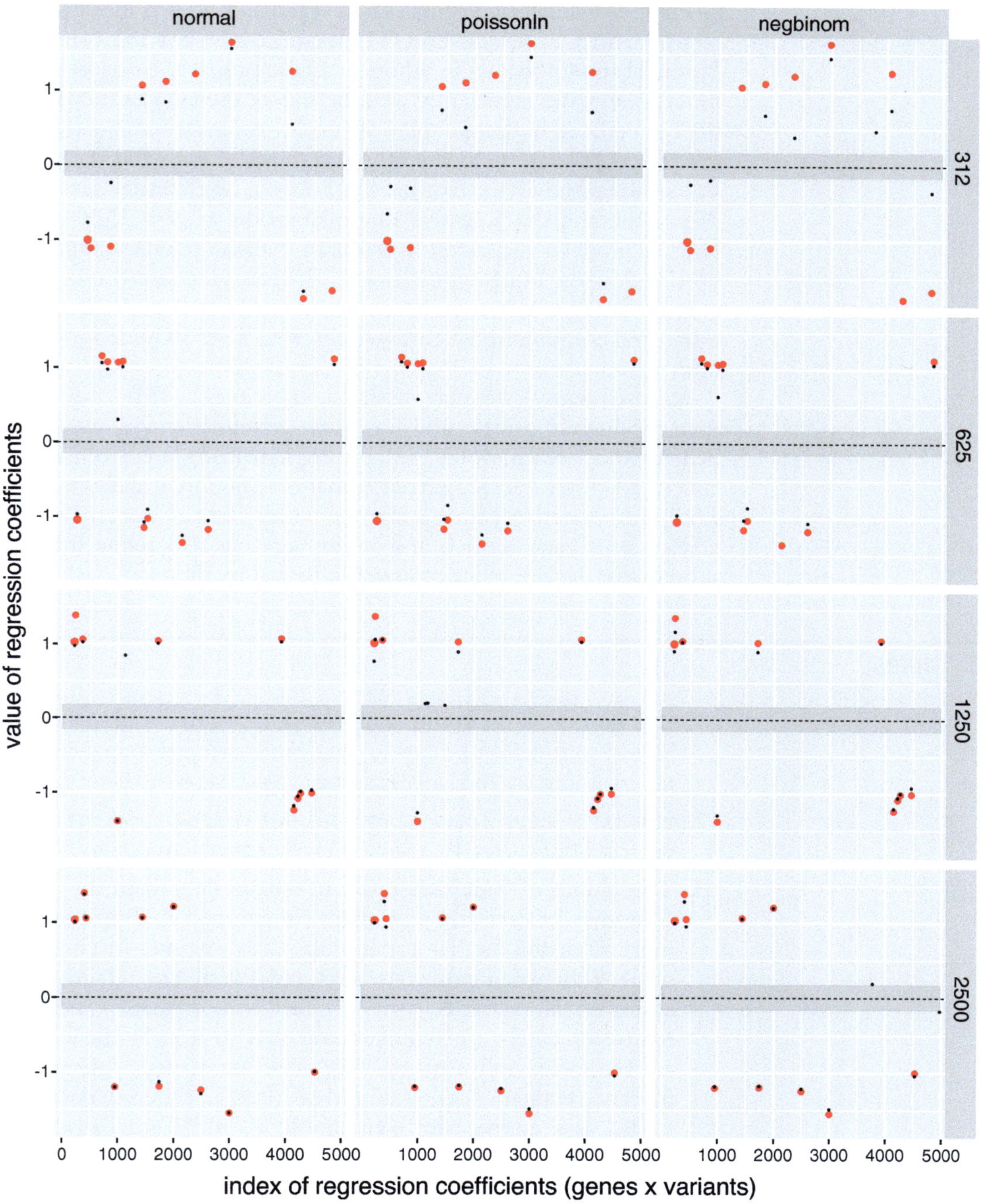

Fig. 3 True (red dots) and estimated (small black dots) regression coefficients for the Normal, Poisson-LogNormal, and Negative Binomial models at four different sample sizes. Regression coefficients within the gray band around zero are considered of negligible size and they are omitted from the plot

Normal, Poisson-LogNormal, and Negative Binomial models correctly identify 7, 8, and 7 eQTLs, respectively, but the Negative Binomial model throws in addition a false positive. None of these models estimates sufficiently well the size of the eQTLs they identify correctly. At higher samples, all models estimate sufficiently well the true eQTLs in the data, although they do not avoid the occasional false positive (see Normal and Poisson-LogNormal models at $N = 1250$ and the Negative Binomial model at $N = 2500$). The Normal model is the fastest to estimate at all sample sizes, followed by the Poisson-LogNormal model. Proper parameterization can greatly facilitate inference in a hierarchical model and reduce its associated computational cost (see the Stan User's Guide for more information).

Subsequently, we may wish to experiment with additional sparse-inducing priors, and/or more systematic and extensive benchmarks, which would require implementing a custom inference algorithm for increased performance. This initial analysis using a fully automated inference system provides a first assessment of the efficiency of each model, and it can serve as a baseline for the subsequent development of novel statistical methods.

5 Notes

1. You can access the command prompt through a terminal emulator. From within `rstudio`, you can create a terminal emulator by going to `Tools>Terminal>New Terminal`.
2. Instead of steps 2 and 3, you can just create a new `rstudio` project (`File>New Project...`) with equivalent results.
3. This chunk of code uses `curl` to stream data from the 1000 Genomes Project and pipe them (in gzipped VCF format) to `vcftools` for further processing. Only variants in chromosome 7 between positions 100 and 200K are retained. In addition, all variants with FILTER other than PASS are removed; indels are removed; variants in non-bi-allelic sites are removed; only variants with allele frequency higher than 5% are retained (i.e., 0.05<MAF<0.95). Filtered data are stored as a matrix with elements 0, 1, and 2 indicating the number of alleles in 496 variants across 2504 samples. If `vcftools` throws an error about not being able to open temporary files, you need to increase the maximum number of open files allowed by your system to more than double the number of samples in the VCF file. At the `bash` command prompt, type: `ulimit -n 5100`. You may need root permissions to run this command.

4. Alternatively, you can create these files through `rstudio` (`File>New File>R Script`).
5. Whenever the code in `utils.R` or `viz.R` changes, these files have to be sourced again to make the changes visible to `analysis.R` by executing lines 2 and 3 of this code chunk. A more structured approach would be to develop an `R` package, which is however beyond the scope of this chapter. The function `doMC::registerDoMC(cores = 8)` registers a parallel backend with 8 cores. This is required for multicore functionality in the `plyr` package.
6. In this and all subsequent code listings, we make heavy use of the pipe operator `%>%` to form long linear chains of functions, where the output of each function becomes the input of the next. This allows for the generation of readable, easy-to-understand workflows. In this particular function, we read genotypic data from disk files in the form of an $M \times K$ matrix. Before returning, the matrix is filtered by removing all variants (i.e., columns) that may have zero variance across all samples (lines 24 and 25). Notice that file names are constructed in a device-independent manner using the function `file.path`. Finally, we prefer the operator = instead of the traditionally used <- for indicating assignment.
7. Data are loaded in the form of an $N \times M$ matrix of read counts. By default, only data for the CEU population are read. If YRI data are preferred, set the function argument to `pop=YRI`, instead. As for the genotypic data, genes with constant expression across all samples are removed (lines 18 and 19).
8. This function implements the *median-of-ratios* method for normalizing a matrix of count data, as presented in [8]. Calculations are performed on the logarithmic scale.
9. The probability mass function of the negative binomial distribution assumed here is parameterized in terms of gene-specific mean μ_i and inverse dispersion α_i parameters, as follows:

$$z_{ij} \sim \frac{\Gamma(z_{ij} + \alpha_i)}{\Gamma(z_{ij} + 1)\Gamma(\alpha_i)} \left(\frac{c_j\mu_i}{c_j\mu_i + \alpha_i}\right)^{z_{ij}} \left(\frac{\alpha_i}{c_j\mu_i + \alpha_i}\right)^{\alpha_i}$$

   where c_j are sample-specific normalization factors.
10. Given an $N \times M$ matrix of counts, we fit the negative binomial distribution to each gene/row, and we estimate gene-specific maximum likelihood estimates of mean and dispersion parameters. For parameter estimation, we use the function `optim`. The code `possibly(optim, otherwise = NULL)` on line 3 returns a version of the `optim` function, which returns `NULL` on error instead of halting execution, thus allowing processing of the whole set of available genes. The parameter `fnscale=-1` on line 11 signals `optim` to perform maximization of the

log-likelihood `lognbinom`, instead of minimization. Iterating over the rows of the data matrix is done using the `plyr` function `alply`, and it is performed in parallel (argument `.parallel=T` on line 15). After all rows have been processed, genes for which estimation failed or did not converge are discarded (lines 16 and 17), results are extracted and properly formatted (lines 18–21), and gene-specific variance values are calculated given the estimated values for mean and dispersion parameters (line 22).

11. An inner function `plot_fn` is defined at the top of `plot_mean_variance`, in order to avoid code repetition when generating plots `gg1` and `gg2` below.
12. This function should be treated as a starting point for more sophisticated data simulation methods. Genotypic data are generated by randomly sampling without replacement `nsamples` rows and `nvars` columns from the observed matrix of genotypes. This approach breaks the spatial correlation between variants, so the variants/columns in the generated matrix X can be treated as being mostly uncorrelated. The matrix B of coefficients is initially generated as a matrix of zeros. In a second stage, `nhits` elements are randomly chosen, and set to non-zero values sampled randomly from a shifted exponential distribution. Half of the `nhits` elements are assigned positive values and the remaining half are assigned negative values. Given matrices X and B and the top `ngenes` estimated mean and dispersion parameters (where parameter values are first ranked in order of decreasing mean values; see line 20), a matrix Z of read counts is generated with elements sampled from the negative binomial distribution. Alternatively, estimated mean and dispersion parameters can be randomly sampled by uncommenting line 19 (and commenting out line 20).
13. The `stan` code presented here is divided in a number of blocks. The `data{...}` block is read exactly once at the beginning of the inference procedure, and it declares required model data. The `transformed data{...}` block in `negbinom.stan` and `poissonln.stan` defines transformations of previously declared data, and it is also executed only once at initialization. The `parameters{...}` block declares model parameters, an unconstrained version of which will be the target of inference procedures. The `transformed parameters{...}` block defines variables in terms of previously defined parameters. All variables that appear in these last two blocks will be returned at the output after inference completes. Finally, the `model{...}` block is where the joint log probability function is defined, using a syntax very close to the actual mathematical notation. Sampling statements (priors) for all variables in the

`parameters{...}` block must appear here, otherwise a uniform, possibly unconstrained, prior is assumed.

14. The argument `auto_write=T` ensures that unnecessary re-compilations of `stan` code are avoided. For this to work, the variables `negbinom`, `poissonln`, and `normal` must reside in the global environment, as in this code listing.
15. `Stan` provides three different inference methodologies [19]: (a) Full Bayesian inference using a self-adjusting Hamiltonian Monte Carlo approach, (b) approximate Bayesian inference using Automatic Differentiation Variational Inference (ADVI), and (c) point parameter estimates by maximizing the joint posterior. Here, for reasons of computational efficiency, we use the third approach (through function `rstan::optimizing`), which however does not provide any measure of uncertainty of the estimates. In principle, such estimates can be derived at a second stage by approximating locally the posterior distribution using the Laplace approximation. `rstan::optimizing` takes as input a compiled model and input data in the form of a list of variables, as they appear in the `data{...}` block of the model definition (variables not in this block are ignored). Function `fit_model` returns the matrices of true and estimated coefficients in vector format, and a vector of the corresponding indices.
16. This is necessary for successfully calling `rstan::optimizing`. Alternatively, we could have imported the whole of `rstan` using `library( rstan)` , and called `optimizing` directly (i.e., without the `::` operator).
17. This code facilitates reproducibility, since for the same seed, the same sequence of pseudo-random numbers is generated.
18. This code chunk includes two nested loops: the outer loop iterates over different sample sizes, while the inner loop iterates over models, and it is executed in parallel. At each iteration of the outer loop, a dataset of appropriate size is simulated and then passed to each tested model in the inner loop.
19. This function takes a `thr` argument: estimated coefficient values below this threshold are considered effectively 0 and set to NA (line 4). When plotting, NA values will be dropped with a warning.

Acknowledgements

This work was supported by the National Institute for Health Research (NIHR) Oxford Biomedical Research Centre Program. The views expressed in this manuscript are not necessarily those of the Department of Health. We also acknowledge the Wellcome Trust Centre for Human Genetics Wellcome Trust Core Award Grant Number 090532/Z/09/Z. The funders of the study had no role in the study design, data collection, data analysis, data interpretation, or writing of the paper.

A File `normal.stan`

```
data {
  int<lower=0> N;                        // number of genes
  int<lower=0> M;                        // number of samples
  int<lower=0> K;                        // number of variants
  matrix[M, K] X;                        // matrix of genotypes
  vector[M] Y[N];                        // matrix of transformed read counts
}

parameters {
  vector[N] b0;                          // baseline gene expression
  vector[N] ls2;                         // log-variance of gene expression
  real<lower=0> eta;                     // global scale parameter
  vector<lower=0>[K] zeta[N];            // local scale parameters
  vector[K] B[N];                        // regression coefficients
}

transformed parameters {
  vector<lower=0>[N] sigma = sqrt(exp(ls2));     // standard deviation of gene expression
}

model {
  real sc = mean(sigma) / sqrt(N*K);
  eta ~ cauchy(0, 1);
  for(i in 1:N) {
    zeta[i] ~ cauchy(0, 1);
    B[i] ~ normal(0, eta * zeta[i] * sc);
    Y[i] ~ normal(b0[i] + X * B[i], sigma[i]);
  }
}
```

B File poissonln.stan

```
data {
  int<lower=0> N;                    // number of genes
  int<lower=0> M;                    // number of samples
  int<lower=0> K;                    // number of variants
  matrix[M, K] X;                    // matrix of genotypes
  int<lower=0> Z[N, M];              // matrix of read counts
  vector<lower=0>[M] c;              // vector of normalisation factors
  vector<lower=0>[M] s;              // vector of library sizes
}

transformed data {
  vector[M] lc = log(c);             // log normalisation factors
  vector[M] ls = log(s);             // log library sizes
}

parameters {
  vector[N] b0;                      // baseline gene expression
  vector[N] ls2;                     // log-variance of gene expression
  real<lower=0> eta;                 // global scale parameter
  vector<lower=0>[K] zeta[N];        // local scale parameters
  vector[K] B[N];                    // regression coefficients
  vector[M] Y[N];                    // latent variables
}

transformed parameters {
  vector<lower=0>[N] sigma = sqrt(exp(ls2));     // standard deviation of gene expression
}

model {
  real sc = mean(sigma) / sqrt(N*K);
  eta ~ cauchy(0, 1);
  for(i in 1:N) {
    zeta[i] ~ cauchy(0, 1);
    B[i] ~ normal(0, eta * zeta[i] * sc);
    Y[i] ~ normal(b0[i] + X * B[i], sigma[i]);
    Z[i] ~ poisson_log(lc + ls + Y[i]);
  }
}
```

C File negbinom.stan

```
data {
  int<lower=0> N;                       // number of genes
  int<lower=0> M;                       // number of samples
  int<lower=0> K;                       // number of variants
  matrix[M, K] X;                       // matrix of genotypes
  int<lower=0> Z[N, M];                 // matrix of read counts
  vector<lower=0>[M] c;                 // vector of normalisation factors
  vector<lower=0>[M] s;                 // vector of library sizes
}

transformed data {
  vector[M] lc = log(c);                // log normalisation factors
  vector[M] ls = log(s);                // log library sizes
}

parameters {
  vector[N] b0;                        // baseline gene expression (log-scale)
  vector[N] lphi;                      // log-dispersion
  real<lower=0> eta;                   // global scale parameter
  vector<lower=0>[K] zeta[N];          // local scale parameters
  vector[K] B[N];                      // regression coefficients
}

transformed parameters {
  vector<lower=0>[N] phi = exp(lphi);            // dispersion
  vector<lower=0>[N] alpha = 1.0 ./ phi;         // inverse-dispersion
}

model {
  real sc = 1.0 / sqrt(N*K);
  eta ~ cauchy(0, 1);
  for(i in 1:N) {
    zeta[i] ~ cauchy(0, 1);
    B[i] ~ normal(0, eta * zeta[i] * sc);
    Z[i] ~ neg_binomial_2_log(ls + lc + b0[i] + X * B[i], alpha[i]);
  }
}
```

D R sessionInfo( )

- R version 3.5.1 (2018–07–02), `x86_64-pc-linux-gnu`
- Locale: `LC_CTYPE=en_GB.UTF-8`, `LC_NUMERIC=C`, `LC_TIME=en_GB.UTF-8`, `LC_COLLATE=en_GB.UTF-8`, `LC_MONETARY=en_GB.UTF-8`, `LC_MESSAGES=en_GB.UTF-8`, `LC_PAPER=en_GB.UTF-8`, `LC_NAME=C`, `LC_ADDRESS=C`, `LC_TELEPHONE=C`, `LC_MEASUREMENT=en_GB.UTF-8`, `LC_IDENTIFICATION=C`
- Running under: `Ubuntu 18.04.1 LTS`
- Matrix products: default
- BLAS: `/usr/lib/x86_64-linux-gnu/openblas/libblas.so.3`
- LAPACK: `/usr/lib/x86_64-linux-gnu/libopenblasp-r0.2.20.so`
- Base packages: base, datasets, graphics, grDevices, methods, stats, utils
- Other packages: bindrcpp 0.2.2, dplyr 0.7.8, forcats 0.3.0, ggplot2 3.1.0, purrr 0.2.5, readr 1.2.1, stringr 1.3.1, tibble 1.4.2, tidyr 0.8.2, tidyverse 1.2.1
- Loaded via a namespace (and not attached): assertthat 0.2.0, backports 1.1.2, base64enc 0.1–3, bindr 0.1.1, broom 0.5.1, callr 3.0.0, cellranger 1.1.0, cli 1.0.1, codetools 0.2–15, colorspace 1.3–2, compiler 3.5.1, cowplot 0.9.3, crayon 1.3.4, doMC 1.3.5, foreach 1.4.4, generics 0.0.2, glue 1.3.0, grid 3.5.1, gridExtra 2.3, gtable 0.2.0, haven 2.0.0, hms 0.4.2, httr 1.3.1, inline 0.3.15, iterators 1.0.10, jsonlite 1.5, labeling 0.3, lattice 0.20–38, lazyeval 0.2.1, loo 2.0.0, lubridate 1.7.4, magrittr 1.5, matrixStats 0.54.0, modelr 0.1.2, munsell 0.5.0, nlme 3.1–137, parallel 3.5.1, pillar 1.3.0, pkgbuild 1.0.2, pkgconfig 2.0.2, plyr 1.8.4, prettyunits 1.0.2, processx 3.2.1, ps 1.2.1, R6 2.3.0, Rcpp 1.0.0, readxl 1.1.0, reshape2 1.4.3, rlang 0.3.0.1, rstan 2.18.2, rstudioapi 0.8, rvest 0.3.2, scales 1.0.0, StanHeaders 2.18.0, stats4 3.5.1, stringi 1.2.4, tidyselect 0.2.5, tools 3.5.1, withr 2.1.2, xml2 1.2.0, yaml 2.2.0

References

1. GTEx Consortium (2013) The genotype-tissue expression (GTEx) project. Nat Genet 45(6):580–585
2. Lappalainen T, Sammeth M, Friedländer MR, AC 't Hoen P, Monlong J, Rivas MA, Gonzàlez-Porta M, Kurbatova N, Griebel T, Ferreira PG, Barann M, Wieland T, Greger L, van Iterson M, Almlöf J, Ribeca P, Pulyakhina I, Esser D, Giger T, Tikhonov A, Sultan M, Bertier G, MacArthur DG, Lek M, Lizano E, Buermans HPJ, Padioleau I, Schwarzmayr T, Karlberg O, Ongen H, Kilpinen H, Beltran S, Gut M, Kahlem K, Amstislavskiy V, Stegle O, Pirinen M, Montgomery SB, Donnelly P,

McCarthy MI, Flicek P, Strom TM, Consortium G, Lehrach H, Schreiber S, Sudbrak R, Carracedo A, Antonarakis SE, Häsler R, Syvänen AC, van Ommen GJ, Brazma A, Meitinger T, Rosenstiel P, Guigó R, Gut IG, Estivill X, Dermitzakis ET (2013) Transcriptome and genome sequencing uncovers functional variation in humans. Nature 501(7468):506–511

3. Vavoulis DV, Taylor JC, Schuh A (2017) Hierarchical probabilistic models for multiple gene/variant associations based on next-generation sequencing data. Bioinformatics 33(19):3058–3064
4. Carpenter B, Gelman A, Hoffman M, Lee D, Goodrich B, Betancourt M, Brubaker M, Guo J, Li P, Riddell A (2017) Stan: a probabilistic programming language. J Stat Softw Articles 76(1):1–32
5. Salvatier J, Wiecki TV, Fonnesbeck C (2016) Probabilistic programming in python using PyMC3. PeerJ Comput Sci 2:e55
6. Polson NG, Scott JG, Clarke B, Severinski C (2012) Shrink globally, act locally: sparse Bayesian regularization and prediction. Oxford University Press, Oxford
7. Carvalho CM, Polson NG, Scott JG (2010) The horseshoe estimator for sparse signals. Biometrika 97(2):465–480
8. Love MI, Huber W, Anders S (2014) Moderated estimation of fold change and dispersion for RNA-seq data with DESeq2. Genome Biol 15(12):550
9. McCarthy DJ, Chen Y, Smyth GK (2012) Differential expression analysis of multifactor RNA-Seq experiments with respect to biological variation. Nucleic Acids Res 40 (10):4288–4297
10. Vavoulis DV, Francescatto M, Heutink P, Gough J (2015) DGEclust: differential expression analysis of clustered count data. Genome Biol 16:39
11. Law CW, Chen Y, Shi W, Smyth GK (2014) Voom: precision weights unlock linear model analysis tools for RNA-seq read counts. Genome Biol 15(2):R29
12. Danecek P, Auton A, Abecasis G, Albers CA, Banks E, DePristo MA, Handsaker RE, Lunter G, Marth GT, Sherry ST, McVean G, Durbin R, 1000 Genomes Project Analysis Group (2011) The variant call format and VCFtools. Bioinformatics 27(15):2156–2158
13. R Core Team R (2018) A language and environment for statistical computing. R Foundation for Statistical Computing, Vienna
14. RStudio Team (2015) RStudio: integrated development environment for R. RStudio, Inc., Boston
15. Wickham H (2011) The split-apply-combine strategy for data analysis. J Stat Softw 40 (1):1–29
16. Wickham H (2007) Reshaping data with the reshape package. J Stat Softw 21(12):1–20
17. 1000 Genomes Project Consortium, Auton A, Brooks LD, Durbin RM, Garrison EP, Kang HM, Korbel JO, Marchini JL, McCarthy S, McVean GA, Abecasis GR (2015) A global reference for human genetic variation. Nature 526(7571):68–74
18. Frazee AC, Langmead B, Leek JT (2011) ReCount: a multi-experiment resource of analysis-ready RNA-seq gene count datasets. BMC Bioinf 12:449
19. Stan Development Team (2018) Stan modeling language user's guide and reference manual, version 2.18.0

Chapter 10

High-Order Association Mapping for Expression Quantitative Trait Loci

Huaizhen Qin, Weiwei Ouyang, and Jinying Zhao

Abstract

Mapping expression quantitative trait loci (eQTLs) is an important avenue to identify putative genetic variants in regulatory regions. Famed eQTL mapping methods exploit the mean effects of locus-wise genetic variants on expression quantitative traits. Despite their successes, such methods are suboptimal because they neglect high-order heterogeneity inherent in genetic variants and covariates. High-order effects of observed loci are common due to their connections to various latent factors, i.e., latent interactions among genes and environmental factors. In this chapter, we introduce a new scheme to harmoniously integrate mean and high-order effects of genetic variants on expression quantitative trait. We rigorously evaluate its validity and utility of signal augmentation.

Key words High-order heterogeneity, Latent gene-by-environment interactions, Latent genetic and nongenetic factors

1 Introduction

Over the past decade, genetic studies on gene expression have enhanced our understanding of transcriptional regulation [1]. Expression quantitative trait loci (eQTL) refer to genomic loci that regulate the expression of mRNA. One key component of eQTL mapping is an association test between genetic loci and expression quantitative traits. The main advantage of eQTL mapping using association methods is the identification of candidate functional loci without requiring the knowledge about cis or trans regulatory regions [2].

Existing prominent eQTL mapping methods merely exploit the mean heterogeneity of a test locus on expression quantitative traits (referred to as MT in context). However, a true eQTL would lead to the alteration of the entire distribution instead of solely the mean of expression quantitative trait. High-order heterogeneity of an eQTL stems from the complicated causal network that involves biological disruption, linkage disequilibrium (LD), gene-by-gene,

Xinghua Mindy Shi (ed.), *eQTL Analysis: Methods and Protocols*, Methods in Molecular Biology, vol. 2082,
https://doi.org/10.1007/978-1-0716-0026-9_10,

and/or gene-by-environment interactions. For example, expression quantitative traits may have different means and variances among different genotypic groups at a specific eQTL [3].

The 4-df versatile likelihood ratio test (LRT) [4] and the joint location-scale (JLS) test [5] would be able to reprioritize eQTLs. The LRT integrates mean and variance heterogeneities of a test locus, assuming normality on trait residual. This integrative method appeared to be more powerful than the MT test in the presence of mean and dispersion effects. The JLS applies Fisher's linear combination of the ordinary MT (i.e., partial t test) on the mean effect of a test locus and Levene's test [6] on variance heterogeneity of trait residuals among genotypic groups. This integrative method appeared more powerful than the MT test in the presence of latent gene-by-environment interactions. However, both methods suffer perceived limitations. Due to the need for an iterative search, the famed LRT is computationally slow. When the normality of trait residual is violated, it may not converge to a meaningless point or may not converge at all. The JLS extends normality of trait residuals to an exponential distribution family. Nevertheless, it requires a very large sample size or large-scale permutation to accurately compute the p value. Of note, both methods are suboptimal in that they require locus-wise genotypic partitions of sample and suffer unnecessarily large degrees of freedom. The two components of each of the famed methods are independent, no matter whether the test locus is a neutral or causal locus.

Here we introduce a new integrative high-order signal augmentation test (HSAT) to overcome these limitations. Unlike the famed methods, the proposed HSAT integrates harmonious primary test and auxiliary test at each test locus. Two tests are called to be harmonious if their test statistics have null independence and alternative dependence. To be specific, null independence requires that two statistics are independently distributed at a neutral locus. Alternative dependence means that the two statistics can be statistically dependent at a true eQTL (and its flanking loci due to linkage disequilibrium). The null independence is necessary and sufficient to properly control type I error rate. The merit of alternative dependence is a major novelty of the HSAT, and this merit is instructive for maximizing statistical power. The HSAT harmoniously integrates mean and high-order heterogeneities by a statistic of two degrees of freedom with no need to partition the sample into different genotypic groups. It computes p value at each locus numerically without any aid of iteration or resampling techniques. In the following sections, we describe this method, mathematically assess its validity, and demonstrate its power gains using published simulation models in supportive of the famed methods.

2 Methods

2.1 The HSAT

Let Υ_i be the residual of expression quantitative trait for subject i after adjusting for various effects of covariates. To account for potential heteroscedasticity, the double generalized linear model (DGLM) [7] would be applied to adjust for both mean and variance effects of data structure and environmental covariates (*see* **Note 1**). In particular, for background population structure, the mean and dispersion effects on the trait should be adjusted for.

At a test locus, let G_i be the copy number of the minor allele for subject i (i.e., $G_i = 0,\ 1,$ or 2). Let the symbols "*mo*", "*ho*," and "*mh*" stand for "modeling the mean effect only," "modeling high-order effect only," and "modeling *m*ean and *h*igh-order effects jointly." The *primary test* assumes the mean-effect model:

$$\Upsilon_i = \mu_1 + G_i\beta_1 + e_i,$$

where μ_1 is the intercept, e_i is random error term, β_1 is the effect size of G_i on Υ_i. The null hypothesis is $H_{0.mo} : \beta_1 = 0$. The primary test statistic is

$$T_1 = \frac{\sqrt{n}\widehat{\sigma}_{\Upsilon,G}}{\sqrt{\widehat{\sigma}^2{}_{\Upsilon}\widehat{\sigma}^2{}_{G} - \widehat{\sigma}^2{}_{\Upsilon,G}}},$$

where $\widehat{\sigma}_{\Upsilon,G} = \frac{1}{n}\sum_{i=1}^{n}(\Upsilon_i - \overline{\Upsilon})(G_i - \overline{G})$, $\widehat{\sigma}^2{}_{\Upsilon} = \frac{1}{n}\sum_{i=1}^{n}(\Upsilon_i - \overline{\Upsilon})^2$, $\widehat{\sigma}^2{}_{G} = \frac{1}{n}\sum_{i=1}^{n}(G_i - \overline{G})^2$, $\overline{\Upsilon} = \frac{1}{n}\sum_{i=1}^{n}\Upsilon_i$, $\overline{G} = \frac{1}{n}\sum_{i=1}^{n}G_i$, and n is sample size. Mathematically, $T_1 = \sqrt{n-2}r_{\Upsilon,G}/\sqrt{1 - r_{\Upsilon,G}{}^2}$, where $r_{\Upsilon,G}$ is the sample Pearson coefficient of correlation between G_i's and Υ_i's.

The *auxiliary test* assumes the secondary model:

$$\Upsilon_i^2 = \mu_2 + G_i^2\beta_2 + e_i',$$

where e_i' is the random error term. The highlight of auxiliary test is to capture the second-order moment information. The null hypothesis of testing the high-order effect is $H_{0.ho} : \beta_2 = 0$. The auxiliary test statistic is

$$T_2 = \frac{\sqrt{n}\widehat{\sigma}_{\Upsilon^2,G^2}}{\sqrt{\widehat{\sigma}^2{}_{\Upsilon^2}\widehat{\sigma}^2{}_{G^2} - \widehat{\sigma}^2{}_{\Upsilon^2,G^2}}},$$

where $\widehat{\sigma}_{\Upsilon^2,G^2} = \frac{1}{n}\sum_{i=1}^{n}\left(\Upsilon_i{}^2 - \overline{\Upsilon^2}\right)\left(G_i{}^2 - \overline{G^2}\right)$, $\widehat{\sigma}^2{}_{\Upsilon^2} = \frac{1}{n}\sum_{i=1}^{n}(\Upsilon_i{}^2 - \overline{\Upsilon^2})^2$, $\overline{\Upsilon^2} = \frac{1}{n}\sum_{i=1}^{n}\Upsilon_i^2$, and $\overline{G^2} = \frac{1}{n}\sum_{i=1}^{n}G_i^2$. Mathematically, $T_2 = \sqrt{n-2}r_{\Upsilon^2,G^2}/\sqrt{1 - r_{\Upsilon^2,G^2}}$, where r_{Υ^2,G^2} is the sample Pearson coefficient of correlation between G_i^2's and Υ_i^2's.

The proposed HSAT integrates T_1 and T_2 to reject the joint null hypothesis $H_{0.mh} : \beta_1 = \beta_2 = 0$. To be specific, the HSAT statistic is defined as $T^2 = T_1{}^2 + T_2{}^2$. Of note, the definitions of sample correlation coefficients and T-statistics do not require normality. For large samples, sample correlation coefficients are consistent estimators for population correlation coefficients, and T statistics are very robust to non-normality. As mathematically proven, at a neutral locus, statistics T_1 and T_2 are asymptotically independent ${\chi_1}^2$ variables and thus $T^2 \to {\chi_2}^2$ under mild moment conditions on the quantitative trait. As such, the p value of T^2 could be computed according to ${\chi_2}^2$ distribution.

2.2 Theoretical Validation

Under the primary model $\Upsilon_i = G_i\beta_1 + e_i$, the covariance between the trait and genotypic score is formulated by $\sigma_{\Upsilon,G} = \sigma_G^2\beta_1$, where σ_G^2 is the variance of the genotypic score. From the primary model, we obtain ${\Upsilon_i}^2 = (G_i\beta_1 + e_i)^2 = \sigma_e^2 + G_i^2\beta_2 + \varepsilon_i$, where σ_e^2 is the variance of random error e_i, $\beta_2 = {\beta_1}^2$ and $\varepsilon_i = 2\beta_1 G_i e_i + {e_i}^2 - \sigma_e^2$. Of note, the newly defined ε_i has mean 0 and variance $\sigma_\varepsilon^2 = 2(1 + 2\beta_1^2\mu_2)\sigma_e^2$. The covariance between Υ_i^2 and G_i^2 is formulated by $\sigma_{\Upsilon^2,G^2} = \beta_2\sigma_{G^2}^2$ and the variance of Υ^2 is $\sigma_{\Upsilon^2}^2 = \sigma_\varepsilon^2 + \beta_2^2\sigma_{G^2}^2$. As such, if G is associated with $\Upsilon(\beta_1 \neq 0)$, then G^2 must be associated with Υ^2. In other words, the mean heterogeneity of a locus alone already implies high-order association, i.e., the association between Υ^2 and G^2. Famed integrative methods neglect this important advantage, whereas the proposed HSAT takes it. Let $E(.)$ denote the expectation of an underlying variable. We formulate the harmonious property of the HSAT as below.

Proposition 1: Under primary model, if $E(e_i^4) < \infty$ and $E(e_i) = E(e_i^3) = 0$, then $T_1 - C_1$ and $T_2 - C_2$ converge to a bivariate normal distribution:

$$\begin{pmatrix} T_1 - C_1 \\ T_2 - C_2 \end{pmatrix} \overset{a.d.}{\rightarrow} \mathcal{N}_2\left(\begin{pmatrix} 0 \\ 0 \end{pmatrix}, \begin{pmatrix} 1 & \rho \\ \rho & 1 \end{pmatrix}\right),$$

where $C_1 = \beta_1(\mu_2 - \mu_1\overline{G})/\sqrt{\widehat{\sigma}_\Upsilon^2\widehat{\sigma}_G^2 - \widehat{\sigma}_{\Upsilon,G}^2}, C_2 = (\beta_1^2(\mu_4 - \overline{G^2}\mu_2) + \left(\mu_2 - \overline{G^2}\right)\sigma_e^3)/\sqrt{\widehat{\sigma}_{\Upsilon^2}^2\widehat{\sigma}_{G^2}^2 - \widehat{\sigma}_{\Upsilon^2,G^2}^2}$, $\rho = \beta_1\delta_1^{-1}\delta_2^{-1}[\sigma_e^2(3\mu_4 - 3\mu_2^2 + \mu_1^2\mu_2 - \mu_1\mu_3) + \beta_1^2(\mu_6 - 2\mu_2\mu_4 + \mu_1\mu_5 - \mu_1^2\mu_4 + \mu_2^3 - \mu_1\mu_2\mu_3 + \mu_1^2\mu_2^2)]$, $\delta_1 = \sqrt{\sigma_e^2\sigma_G^2}$, $\delta_2 = \sqrt{2(1 + 2\beta_1^2\mu_2)\sigma_e^2\sigma_{G^2}^2}$, and σ_e^2, σ_G^2, and $\sigma_{G^2}^2$, are the variances of e, G, and G^2, respectively.

Explicitly, Proposition 1 formulates the harmonious property of the statistics T_1 and T_2 under the primary model with very mild moment conditions (*see* **Note 2**). If the trait is independent of the test locus ($\beta_1 = 0$), then $C_1 = C_2 = \rho = 0$, and hence T_1 and T_2 asymptotically follow the standard bivariate normal distribution $\mathcal{N}_2(\mathbf{0}, \boldsymbol{I}_2)$. With the joint normality, zero correlation ($\rho = 0$) warrants the independence between statistics T_1 and T_2 under the null hypothesis. On the other hand, if the trait of interest is associated to the test locus ($\beta_1 \neq 0$), then T_1 and T_2 turn to be asymptotically dependent ($\rho \neq 0$). As such, the HSAT is more powerful than the LRT and JLS methods to identify a locus that only has mean effect on the trait. For such a locus, the LRT and JLS purely waste 2 degrees of freedom. In practice, high-order association can be driven by other factors, e.g., latent dispersion effects and latent gene-by-environment interactions.

2.3 Empirical Demonstration

2.3.1 Dispersion Design

For empirical comparisons, we first generated individual trait values and genotypes from $\Upsilon = -0.03X_1 - 0.08X_2 + 0.23[X_0 + \delta(X_1 + 1.16X_2)]\varepsilon$, where X_k is indicator of event "$G = k$" for $k = 0,\ 1,\ 2$, the genotypic score G follows the binomial distribution $\mathcal{B}\text{inom}(2, \text{MAF})$, MAF is the minor allele frequency, residual ε follows the standard normal distribution $\mathcal{N}(0, 1)$. This data-generating model is exactly the dispersion model in [4] when $\delta = 25/23$ and $\text{MAF} = 0.4$. By ref. 8, variants with $\text{MAF} < 1/\sqrt{2n}$ are taken as rare variants. In our simulations, we set $\text{MAF} = 0.02$ and sample size $n = 1000$. For power comparisons, we varied the dispersion coefficient δ from 1 to 2. To inspect type I error rate control, we generated individual trait values from null dispersion model $\Upsilon = 0.23\varepsilon$.

2.3.2 Latent Interaction Design

Following Soave et al. [5], we generated individual values of quantitative trait, genotype of a causal locus, and latent exposure from $\Upsilon = G\beta_G + GE\beta_{GE} + e$, where genotypic score G follows binomial distribution $\mathcal{B}\text{inom}(2, \text{MAF})$, MAF is the minor allele frequency, the latent exposure follows Bernoulli $\mathcal{B}(1, 0.3)$, and trait residual e follows the standard normal distribution $\mathcal{N}(0, 1)$. In our simulations, we set $\text{MAF} = 0.3$ and sample size $n = 1000$. The model collapses to the null when setting $\beta_G = \beta_{GE} = 0$ to investigate type I error rate control. For power comparison, we set $\beta_G = 0.01$ and varied the interaction effect β_{GE} from 0 to 1 by a grid of 0.1. Under this design, individual values of the latent exposure and the interaction terms were hidden when computing the p values of all the tests.

2.3.3 Type I Error Rate Control

By large-scale simulations under the null models, our HSAT and the competitors appeared properly and controlled type I error rates at genome-wide significance level (5×10^{-8}). Table 1 demonstrates their rigorous validity. Under each null model, the large number of p values of each test followed $\mathcal{U}(0, 1)$, granting the validity of the test. The rigorous validity of the proposed HSAT stems from the null independence between its two components, namely, they are independently distributed when the trait is independent of the test locus. Besides, our HSAT accurately computes the p values by using very robust numerical algorithms.

Table 1
Type I error rates control

Method	Null dispersion model		Null interaction model	
	Error rate[a]	$P_{K\text{-}S}$[b]	Error rate[a]	$P_{K\text{-}S}$[b]
MT	4×10^{-8}	0.9946	6×10^{-8}	0.9632
JLS	2×10^{-8}	0.8756	4×10^{-8}	0.8258
LRT	3×10^{-8}	0.9567	1×10^{-8}	0.1594
HSAT	5×10^{-8}	0.9238	3×10^{-8}	0.8136

[a]Under each null model, the type I error rate of each test was computed as the proportion 10^8 null p values of $<5 \times 10^{-8}$
[b]p-Value of the Kolmogorov–Smirnov test on the null hypothesis that the 10^8 null p values follow the standard uniform distribution $\mathcal{U}(0, 1)$

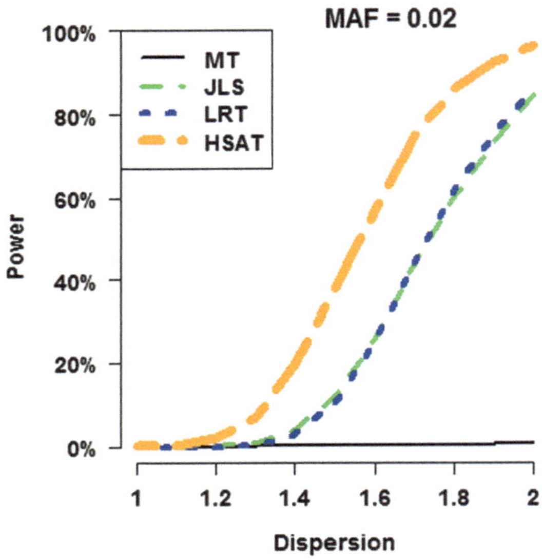

Fig. 1 Power comparison under dispersion model with rare variant. Each point on each power curve was calculated from 10^6 replicates of 1000 simulated subjects from the specific model setup at significance level 5×10^{-8}

2.3.4 Power Comparisons

By large-scale simulations, the HSAT appeared significantly more powerful than all three competitors, where the LRT and JLS appeared equally powerful and both outperformed the classical MT. Under the dispersion model (Fig. 1), the power gain of the HSAT over the famed LRT and JLS reached about 40% when dispersion coefficient δ was around 1.6. Of note, the power gains appeared especially striking for identifying rare variants since the sample partitions in the LRT and JLS reduce power.

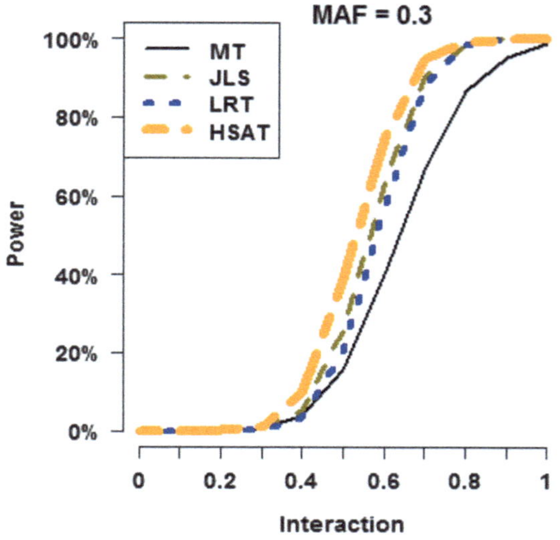

Fig. 2 Power comparison under interaction model with a latent exposure. Each point on each power curve was calculated from 10^6 replicates of 1000 simulated subjects from the specific model setup at significance level 5×10^{-8}

Under the latent interaction design, neither individual values of the latent exposure nor the interaction terms were directly modeled to compute the p values. The three integrative methods captured the interaction effects via high-order heterogeneity of the genotypic score in different ways, whereas the ordinary MT simply ignored the latent interaction effect. When interaction effect β_{GE} was around 0.5, the power gain of the HSAT over the MT, LRT, and JLS reached about 25%, 20%, and 15%, respectively (Fig. 2). The HSAT appeared more powerful than the competitors uniformly, whereas the LRT and JLS outperformed the ordinary MT when interaction effect was strong enough (e.g., $\beta_{GE} > 0.4$). The famed LRT and JLS appeared slightly less powerful than the ordinary MT when the interaction effect was mild (e.g., $\beta_{GE} < 0.4$).

The HSAT overcame the deficiencies of the sample partition in the integrative LRT and JLS. Its uniform optimality and significant power gains stemmed from the dependence between its two components. They reinforced each other in the presence of high-order heterogeneities as driven by the dispersion and latent interaction effects of causal loci (*see* **Note 3**).

2.4 Implementation in R

In the following R function HSAT, the "geno" is the $n \times 1$ vector of genotypic scores and "traitRes" is the $n \times 1$ vector of trait residuals after adjusting for data structure and environmental covariates (*see* **Note 1**). This function returns the overall p value after harmoniously integrating the mean and high-order heterogeneities of a single test locus. This function is highly efficient in computation because it only calls the efficient correlation test function cor.

test (.) and the pchi2(.) built in the R base package. It computes and combines two t-tests and evaluates the overall p value without any iterative or resampling techniques.

```
HSAT<-function(geno, traitRes, covariate=NULL)
{
x<-as.numeric(geno)
y<-as.numeric(traitRes)
T1<-cor.test(x, y)[1]
T2<-cor.test(x**2, y**2)[1]
Pval<-pchisq(q=T1**2+T2**2, df = 2, lower.tail=FALSE)
return (Pval)
}
```

3 Notes

1. Before applying the HSAT method in eQTL mapping, it is necessary to effectively calibrate mean and variance effects of data structure of environment covariates such as age and gender. The conventional linear adjustment only adjust for the mean effects of covariates and is often insufficient. The double generalized linear model (DGLM) [7] would work.
2. The validity of the proposed HSAT is warranted by large-sample theory (Proposition 1). The joint asymptotic normality of its two test statistics does not require normality of the error terms. It only requires $E(e_i^4) < \infty$ and $E(e_i) = E(e_i^3) = 0$. These moment conditions are very mild and cover a broad range of symmetric and asymmetric error distributions.
3. The HSAT method outperforms existing integrative mean-variance methods such as JLS and LRT in terms of both power and computation efficiency. It combines primary test and high-order auxiliary test without partitioning sample into subgroups. Meanwhile, its mean and high-order components reinforce each other when the test variant is truly associated with the trait of interest. It numerically computes p values without using any resampling techniques. As such, it is an effective and efficient avenue for mapping eQTL associations.

Acknowledgments

This work was partially funded by the National Institutes of Health grants R01DK091369, R01MH097018, and RF1AG052476. The funders had no role in study design, data analysis, preparation of the manuscript, or decision to publish.

References

1. Rockman MV, Kruglyak L (2006) Genetics of global gene expression. Nat Rev Genet 7:862–872
2. Nica AC, Dermitzakis ET (2013) Expression quantitative trait loci: present and future. Philos Trans R Soc B 368:20120362
3. Hulse AM, Cai JJ (2013) Genetic variants contribute to gene expression variability in humans. Genetics 193:95–108
4. Cao Y, Wei P, Bailey M, Kauwe JS, Maxwell TJ (2014) A versatile omnibus test for detecting mean and variance heterogeneity. Genet Epidemiol 38:51–59
5. Soave D, Corvol H, Panjwani N, Gong J, Li W, Boëlle P-Y, Durie PR, Paterson AD, Rommens JM, Strug LJ (2015) A joint location-scale test improves power to detect associated SNPs, gene sets, and pathways. Am J Hum Genet 97:125–138
6. Levene H (1960) Robust tests for equality of variances. In: Olkin I (ed.) Contributions to probability and statistics, vol 1. Stanford University Press, Palo Alto, pp 278–292
7. Smyth GK, Verbyla AP (1999) Adjusted likelihood methods for modelling dispersion in generalized linear models. Environment 10:695–709
8. Tony Cai T, Jessie Jeng X, Jin J (2011) Optimal detection of heterogeneous and heteroscedastic mixtures. J R Stat Soc Series B Stat Methodol 73:629–662

Chapter 11

Integration of Multi-omics Data for Expression Quantitative Trait Loci (eQTL) Analysis and eQTL Epistasis

Mingon Kang and Jean Gao

Abstract

Expression quantitative trait loci (eQTL) mapping studies identify genetic loci that regulate gene expression. eQTL mapping studies can capture gene regulatory interactions and provide insight into the genetic mechanism of biological systems. Recently, the integration of multi-omics data, such as single-nucleotide polymorphisms (SNPs), copy number variations (CNVs), DNA methylation, and gene expression, plays an important role in elucidating complex biological systems, since biological systems involve a sequence of complex interactions between various biological processes. This chapter introduces multi-omics data that have been used in many eQTL studies and integrative methodologies that incorporate multi-omics data for eQTL studies. Furthermore, we describe a statistical approach that can detect nonlinear causal relationships between eQTLs, called eQTL epistasis, and its importance.

Key words eQTL mapping study, Integrative analysis, Multi-omics, eQTL epistasis, Single-nucleotide polymorphism, Gene expression, Copy number variation, DNA methylation

1 Introduction

Understanding the genetic casual effects and interactions of complex biological processes has been a central issue in biomedical research. Expression quantitative trait loci (eQTL) studies have played a key role in unveiling genetic regulatory mechanisms of complex biological systems as well as Genome-Wide Association Studies (GWAS). eQTL studies identify genetic loci that regulate gene expression. Genes determine protein's functionality in cells, and its intermediate position of gene expression between genotype and phenotype allows an in-depth exploration bridging the gap between them.

Most eQTL studies have examined genetic variations of single-nucleotide polymorphisms (SNPs) associated to gene expression. However, recently the importance of multi-omics data has been increasingly highlighted for a better understanding of the complex biological systems. This chapter aims to introduce multi-omics data

Xinghua Mindy Shi (ed.), *eQTL Analysis: Methods and Protocols*, Methods in Molecular Biology, vol. 2082,
https://doi.org/10.1007/978-1-0716-0026-9_11,

of single-nucleotide polymorphisms, copy number variations, DNA methylation, and gene expression, as well as integrative methodologies for multi-omics eQTL analysis. Moreover, it describes eQTL epistasis, which is a nonlinear interaction between eQTLs. eQTL epistasis can provide a solution to infer hierarchal relationships between genes, so that gene regulatory network can be reconstructed.

2 Materials

2.1 Single-Nucleotide Polymorphisms (SNPs)

Single-nucleotide polymorphism (SNP pronounced "snips") is a DNA variation in a single nucleotide which is designated by statistical criteria within a population (e.g., >1%) [1]. The human genome contains approximately three billion DNA base pairs, which reside in the 23 pairs of chromosomes. Since 99.9% of human genomes are identical in every person, the investigation of the DNA variations that differ in a population is more efficient than examining every DNA base pairs. SNPs are typically considered as potential genetic biomarkers that may cause diseases.

Many SNPs are found between genes in DNA, which are noncoding regions, more frequently than in coding regions. When SNPs occur in regulatory regions such as promoters and enhancers of a gene, they may regulate the gene and eventually play an important role in phenotypic variations via proteins that the gene encodes.

Most SNPs have only two alleles, where one is major allele and another is minor allele. The major allele is the most common allele in a given population, and the minor allele is the second most one. Minor allele frequency (MAF) refers to the frequency where minor allele occurs. Each SNP has two nucleotides, one from each chromosome. Thus, the possible genotypes are *aa*, *Aa*, and *AA*, where *a* represents a minor allele and *A* represents a major allele in a SNP. It is called a homozygous SNP (e.g., *aa* or *AA*) if both genotypes from each chromosome carry the same DNA variations, a heterozygous SNP (e.g., *Aa*) if having different mutations between the two genotypes in an individual. For instance, Fig. 1a presents an example of SNPs, where a row shows SNPs of an individual and a column shows nucleotides on a SNP locus across the samples. A bold font implies major alleles, while a normal font represents minor alleles in the figure. For instance, the DNA base of "G" is a major allele whereas "A" is a minor allele on SNP1 in Fig. 1a.

The genotype effects for SNPs may be related to the number of minor alleles shown in a small population rather than major alleles that most people have. Phenotypes (we mainly consider human diseases in this chapter) are directly or indirectly associated with the given genotype effects. Standard models of the relationship between genotype and phenotype are mainly threefold: (1) additive,

	SNP1		SNP2		SNP3		SNP4	
Sample 1	**G**	A	**A**	C	**C**	T	**C**	T
Sample 2	**G**	**G**	**A**	**A**	**C**	**C**	**C**	**C**
Sample 3	**G**	**G**	**A**	**A**	**C**	**C**	**C**	**C**
Sample 4	**G**	A	**A**	C	**C**	**C**	**C**	**C**
Sample 5	A	A	C	C	T	T	T	T

A. SNP Genotypes

Encoding ⇒

	SNP1	SNP2	SNP3	SNP4
Sample 1	1	1	1	1
Sample 2	0	0	0	0
Sample 3	0	0	0	0
Sample 4	1	1	0	0
Sample 5	2	2	2	2

B. Additive mode

	SNP1	SNP2	SNP3	SNP4
Sample 1	1	1	1	1
Sample 2	0	0	0	0
Sample 3	0	0	0	0
Sample 4	1	1	0	0
Sample 5	1	1	1	1

C. Dominant mode

	SNP1	SNP2	SNP3	SNP4
Sample 1	0	0	0	0
Sample 2	0	0	0	0
Sample 3	0	0	0	0
Sample 4	0	0	0	0
Sample 5	1	1	1	1

D. Recessive mode

Fig. 1 Encoding SNPs in eQTL analysis. Example of the three encoding modes of SNPs in eQTL analysis: (**b**) Additive mode, (**c**) Dominant mode, and (**d**) Recessive mode for (**a**) for original SNP genotypes

(2) dominant, and (3) recessive models. The additive model assumes that the risk of the disease may be proportional to the number of minor alleles, i.e., *aa* may cause double risks of the disease than *Aa* in the additive manner. The genotypes are encoded as $aa \rightarrow 2$, $Aa \rightarrow 1$, and $AA \rightarrow 0$ as shown in Fig. 1b. The dominant model implies the presence of minor allele may increase the disease risk (Fig. 1c). *aa* has identical genotype effect with *Aa* in the dominant manner ($aa \rightarrow 1$, $Aa \rightarrow 1$, and $AA \rightarrow 0$). In the recessive model (Fig. 1d), two minor alleles (*aa*) are required for the disease risk ($aa \rightarrow 1$, $Aa \rightarrow 0$, and $AA \rightarrow 0$). The three models (Fig. 1b–d) are used for SNP genotype encoding and further analysis.

2.2 Gene Expressions

DNA makes RNA in transcription, and RNA produces proteins dictating cell function in translation. In other words, protein production begins at transcription, so the process of RNA transcripts determines protein's functionality in cells in biological systems [2–4]. The intermediate position of gene expression between genotype and phenotype makes it possible to capture the insight of the genetic architecture of gene expression that bridges the gap between them.

The advance of high-throughput microarray technologies allows one to measure thousands of gene expressions simultaneously [2, 5]. Advances coming with its advantages of the high-throughput techniques enable to obtain increasing numbers of samples, and consequently promising studies of quantitative analysis as well as qualitative analysis become possible. Gene expression data is actively used in many biomedical studies such as gene regulatory network, diseases classification/prediction, drug sensitivity, and identification of disease-specific biomarkers.

The recent development of next-generation sequencing (NGS) techniques, RNA-seq (RNA sequencing) is replacing gene expression microarrays [6]. Along with the trend of NGS, many eQTL studies currently use RNA-seq data. eQTL studies using RNA-seq allows one to identify allele-specific expression (ASE) and isoform gene expression [7].

2.3 Multi-omics Data

Human diseases are caused by a sequence of complex interactions between multiple biological processes, such as genetic, epigenetic, transcriptional, and posttranscriptional regulations [8–12]. For instance, monozygotic twins with identical DNA sequences often have different copy number variations, and it consequently causes discordance in congenital heart defects [13, 14]. Therefore, the consideration of the multiple types of biological processes is essential in dissecting complex biological systems.

The rapid development of high-throughput DNA sequencing and microarray technologies makes efficient data acquisitions possible and provides global snapshots of multi-biological processes. Moreover, continuous improvements and advances in emerging technologies have been expanding available genomic data types, while reducing costs. Multi-omics data such as single-nucleotide polymorphism, copy number variation, DNA methylation, micro-RNA, noncoding RNA, and gene/protein expression are available for measuring the states of the multiple biological processes. In eQTL analysis, copy number variations, DNA methylations, and gene expressions are widely used [15–19].

Copy number variations (CNVs) are modified gene structures, where specific regions of the genome are deleted or duplicated on a chromosome. CNVs are strongly associated to the expressions of genes mapped within or nearby the CNVs, and affect alteration of gene dosage, downstream pathways, and regulatory pathways [20, 21], although CNV is frequently observed even in healthy individuals. The associations between CNVs and diseases have been reported through a number of whole-genome association studies [21]. The deletion (or insertion) of either particular region within a gene or a regulatory region of a gene may result in a lower (or higher) gene expression than what is normally expressed (vice versa).

DNA methylations (DMs) are epigenetic regulatory modifications. DMs often inhibit gene expressions in a nearby gene by adding methyl groups to cytosine (C) or adenine (A) of DNA [22, 23]. It is reported that most CpG islands in mice are covered methylation maps [24]. High levels of 5-methylcytosine in the promoter region of the nearby gene play an important role in gene expression levels even on the same DNA sequence.

Since much research has considered SNPs, CNVs, DNA methylations, and gene expressions for integrative studies of multi-omics data, we focus on introducing integrative eQTL analysis using those multi-omics data in this chapter.

3 Methods

3.1 Genome-Wide Association Studies (GWAS)

Genome-wide association studies (GWAS) examine associations between common genetic variations across the entire human genome and observable traits (e.g., human diseases). GWAS typically perform statistical tests for identifying statistically significant SNPs in the two groups of case and control, where the null hypothesis is that there is no genotypic difference between the groups [25].

Nearby SNPs are often inherited together on the same chromosome, and the SNPs tend to be highly correlated, which refers to linkage disequilibrium (LD) [26]. The highly correlated SNPs make it difficult to compute statistical power analytically for the association tests. The problem refers to multicollinearity in statistics, which occur when more than two variables are highly correlated. To tackle the problem, the highly correlated SNPs are used to be represented as a single value. For instance, the average number of minor alleles of the highly correlated SNPs can represent the genetic effect of the loci. Alternatively, a single SNP can be considered as the representative removing its highly correlated other neighbor SNPs.

For identifying disease-specific SNPs, most genome-wide association studies compute single-locus statistic tests by examining each SNP associated to a phenotype independently. The statistic association tests include analysis of variance (ANOVA), logistic regression, chi-square test, and Fisher's exact test (*see* **Note 1**). Moreover, linear-model-based feature selection approaches have been adapted for GWAS analysis [27, 28]. Elastic-net regularization on linear regression models provides a sparse solution when data has multicollinearity. In the study [27], a stepwise solution identifies disease-causing SNPs.

GWAS often introduce high false-positives due to a large number of simultaneous statistical tests for multiple comparisons (called a multiple testing problem). Statistical tests determine the significance of SNPs with a p value comparing to a cutoff value (normally $\alpha = 0.05$). In GWAS, more than 10^6 numbers of statistical tests are performed on human SNPs. If the SNPs on the test are independent, the false-positive would be $\alpha \times 10^6 = 50{,}000$ when $\alpha = 0.05$. Therefore, the correction of the cutoff is necessary for the stringent control of false-positive errors. Bonferroni's correction adjusts the threshold from 0.05 to $0.05/10^6 = 5 \times 10^{-8}$ [29]. A Manhattan plot visualizes the significance of SNPs, where the negative logarithm of p values for each SNP is plotted.

3.2 eQTL Mapping Study

Identifying polymorphic regulatory regions that control gene expression is essential for unveiling gene regulatory mechanisms [30–32]. Expression quantitative trait loci (eQTL) mapping studies explore genetic loci that regulate gene expression, while GWAS examine associations between genetic variations and phenotype.

Like GWAS, eQTL mapping studies also have the multicollinearity problem on gene expression data due to the characteristics of coexpression, coregulation, and cofunctionality of neighboring genes [33]. A number of researches have attempted to tackle the multicollinearity problem by using machine learning techniques, such as clustering techniques, hierarchal models, partitioning, and pathway based methods. The gene expression data clustered by clustering techniques such as spectral clustering and hierarchical clustering was introduced for further eQTL analysis with sparse partial least squares (sPLS) [34] or a multi-task LASSO model [35]. Gene expression and SNPs were partitioned as modules and the associations were identified by a Markov chain Monte Carlo algorithm [36].

eQTLs are often considered as cis- and trans-eQTLs according to the relative locations of the eQTLs and the genes that they influence. Cis-eQTLs are located in enhancers and promoters of a gene which directly control the expression level of the gene [30, 37]. The cis-acting regulatory regions can be defined as certain upstream regions of a gene in the same chromosome. The boundary ranges for cis-eQTL are often set from 5 Mb up to 20 Mb. In contrast, trans-eQTLs influence a gene at a distance, even often in other chromosomes.

Most eQTL studies generate much lower number of samples comparing to the number of features such as SNPs and gene expression. About a million of SNPs and more than twenty thousand gene expressions are available while most studies consider hundreds of samples, often causing high false-positive errors in the analysis. Feature selection preprocessing and multivariate based methods may be the potential solutions for this problem [38].

3.3 Multi-omics eQTL Analysis

The importance of integrative studies has been increasing in an emerging era of various high-throughput multi-omics data [9–11, 39] (*see* **Note 2**). Recently multi-omics data have been widely incorporated in a large number of research projects to decipher

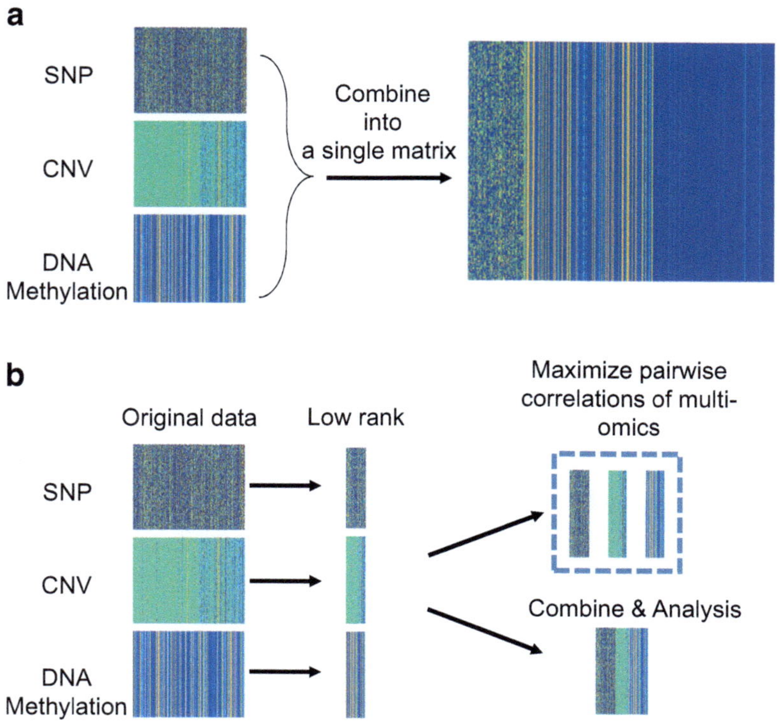

Fig. 2 (**a**) Combination of multi-omics data into a single matrix and (**b**) latent variable mode base integration

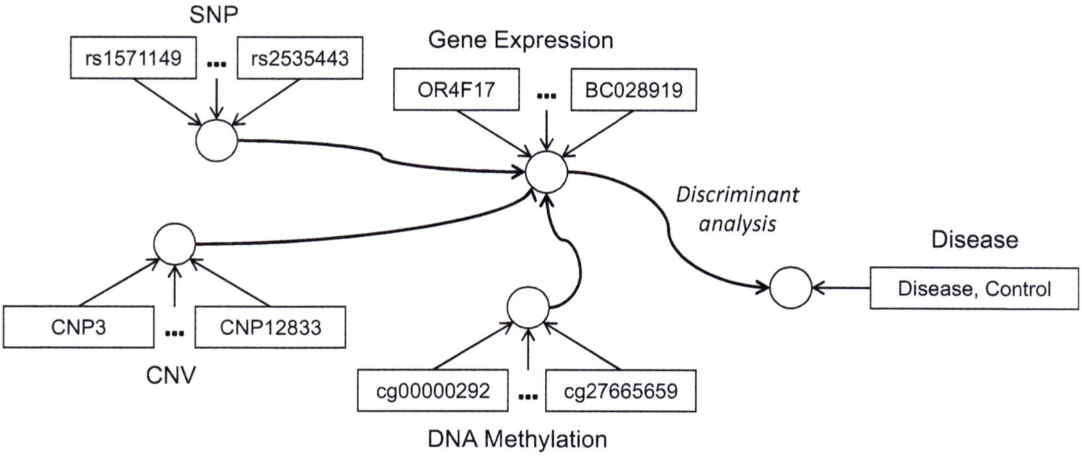

Fig. 3 Multi-omics discriminant analysis [17]

the complex mechanism of human diseases rather than using a single type of genomic data that most genomic research traditionally have done [40, 41]. The integrative approaches of multi-omics data are mainly bifold: (1) augmentation of multi-omics data into a single matrix and (2) latent variable model-based approaches.

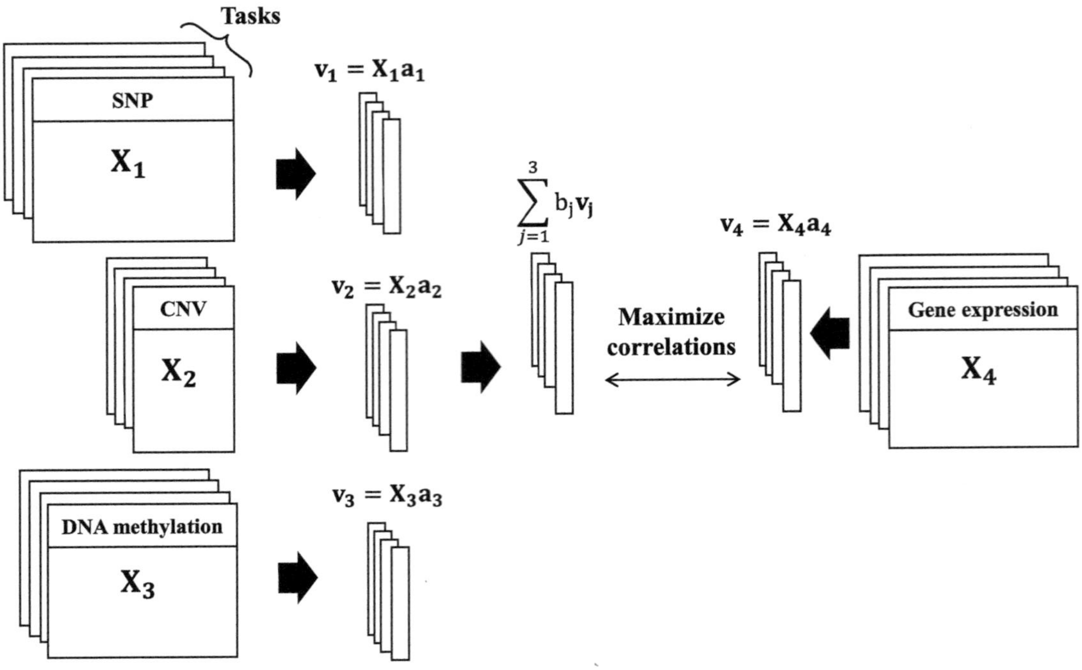

Fig. 4 Canonical correlation analysis for multi-omics data and multi-labeled diseases [43]

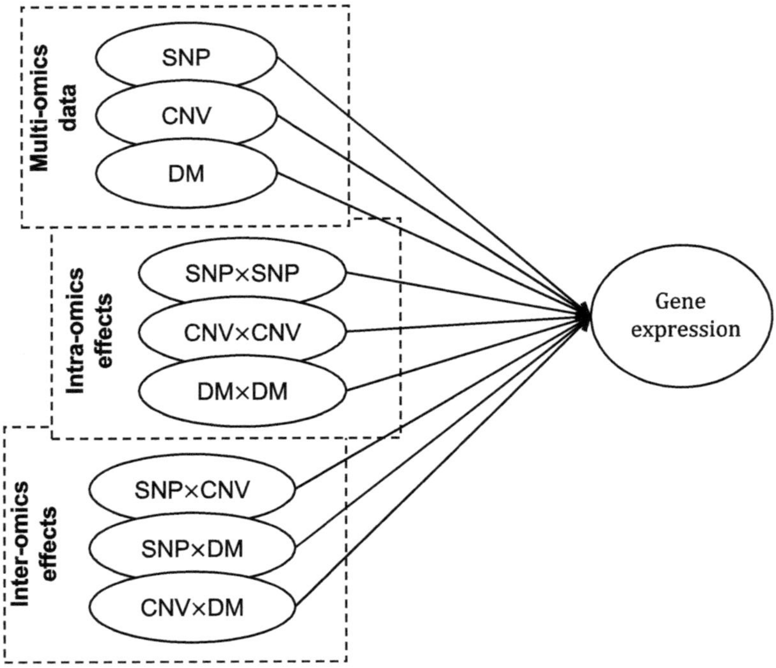

Fig. 5 Bipartite graph for intra/inter-interaction of multi-omics [19]

The simplest approach that integrates multi-omics data is to combine all types of genomic data into a single matrix (Fig. 2a). Then, the augmented matrix that represent the multi-omics data is introduced to statistical methods for further analysis. The multi-omics data of DNA methylation, CNV, miRNA expression, and gene expression were combined into an augmented matrix, and Cox regression was performed for identifying ovarian cancer subtypes [41]. Similarly, ovarian cancer data of DNA methylation, protein, miRNA, and gene expression were classified by a Cox regression model and identified combinational effects of multi-omics data [42].

The alternative approaches are based on latent variable models (Fig. 2b). In the approaches, each omics dataset is considered as an independent block; the blocks are transformed to the lower dimensional data preserving the original information. Specifically, sparse generalized canonical correlation-based approaches (SGCCA) represent each omics data as a block and identify their subsets that maximize the total correlations. Multi-block discriminant analysis (MultiDA) identifies multi-omics components of the highest correlation with the gene sets that discriminate diseases from control [17]. Figure 3 shows the graphical representation of MultiDA. The total three pairwise correlations between {SNP, gene expression}, {CNV, gene expression}, and {DNA methylation, gene expressions} are maximized in MultiDA, while exposing more weights to discriminative genes for disease prediction.

The multi-block and multi-task learning (MBMT) method is developed based on the sparse generalized canonical analysis [43]. Similar to MultiDA, MBMT takes the low dimensional subspace of each omics data. However, it maximizes the correlation between the linear combination of latent variables of SNPs, CNVs, and DNA methylations and the subspace of gene expression (Fig. 4). It enables to identify the combinatorial effects of multi-omics data. Moreover, MBMT considers different functionalities of multi-omics data on the multiple diseases in the model by adapting multitask learning. For instance, psychiatric disorders, such as bipolar disorder, schizophrenia, and major depression, share most biological components, but have different effect sizes on them. Multitask learning identifies common subcomponents in the systems, but having different coefficients. In Fig. 4, $\mathbf{X}_1 - \mathbf{X}_4$ represent SNP, CNV, DNA methylation, and gene expression, respectively, and each data has multiple samples of different disease classes (three psychiatric disorders and control in the study). $\mathbf{v}_1 - \mathbf{v}_4$ show the latent variable of each omics data. The maximized correlation between the linear combination of $\mathbf{v}_1 - \mathbf{v}_3$ and the latent variable of gene expression $\mathbf{v}_4$ is identified for the analysis.

Nonnegative matrix factorization (NMF) has been widely used in many integrative studies. The multi-omics data of SNP, CNV, DNA methylation, gene expression, and miRNA expression were projected into low dimensional matrices by a joint NMF and combined linearly. The combinational subsets of multi-omics data, which are highly correlated with functional associations of ovarian cancer, were reported to discern the difference of patient subgroups [44]. An integrative association study also incorporated multi-omics data by NMF with a sparsity option to identify significant multi-modal genomic components [45].

Both sparse generalized canonical correlation analysis and non-negative matrix factorization approaches assume that each omics data are independent. However, most biological systems involve complex interactions in multiple biological processes rather than a single operation [19, 46]. Multi-block bipartite graph (MB2I) introduces intra/inter-omics effects between multi-omics data in a bipartite graph [19], where eQTL mapping structures of multi-omics data are represented by the bipartite graph. A bipartite graph consists of two disjoint sets (**U** and **V**) and edges that connects between the two sets but do not within each set. For multi-omics eQTL analysis, the set **U** contains multi-omics data of SNPs, CNVs, DNA methylations and **V** includes gene expressions. The edges between the sets show their associations. Specifically, the set **U** in the bipartite graph includes the three major effect groups, which are main effect of the multi-omics data, intra-omics effect, and inter-omics effect, as illustrated in Fig. 5. The main effect accounts for individual main effect of each omics data themselves; the intra-omics effect shows the interaction effect between the features in the same data type; and the inter-omics effect is the interaction between the features across the different types. Statistically significant associations between the two sets are inferred. The graph-based multi-omics eQTL analysis can take advantage of the multitude of graph analysis tools/algorithms such as maximum edge biclique and maximum spanning tree, while providing intuitive insight of biological systems.

In the integrative analysis with multi-omics data, simulation study is challenging [47]. The ground truth of well-known biological system does not exist in most complex human diseases, so most research conducts simulation study to assess their proposed methods. However, the generation of simulation data for multi-omics data is very difficult to implement due to their high complexity and complex interactions.

Binding sites of DNA-associated protein can be accurately identified by chromatin immunoprecipitation (ChIP) with massively parallel DNA sequencing (ChIP-seq) or ChIP followed by microarray hybridization (ChIP-chip) [48]. Especially, ChIP-seq techniques provide functional genomic analysis at higher resolution than DNA microarrays. ChIP-seq analysis makes high-throughput cis-eQTL analysis possible.

A

Multiplicative model
Within/between loci

	BB	Bb	bb
AA	α	αr_2	αr_2^2
Aa	αr_1	$\alpha r_1 r_2$	$\alpha r_1 r_2^2$
aa	αr_1^2	$\alpha r_1^2 r_2$	$\alpha r_1^2 r_2^2$

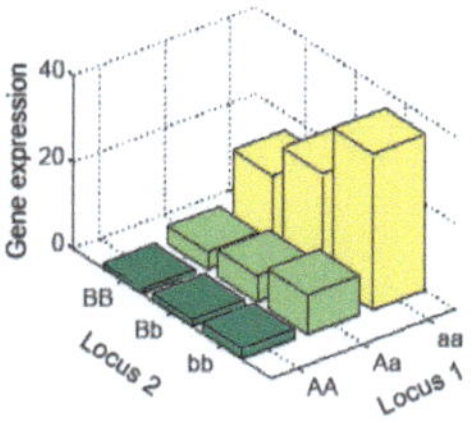

B

Multiplicative model
with threshold
Within/between loci

	BB	Bb	bb
AA	α	αr_2	αr_2
Aa	αr_1	$\alpha r_1 r_2$	$\alpha r_1 r_2$
aa	αr_1	$\alpha r_1 r_2$	$\alpha r_1 r_2$

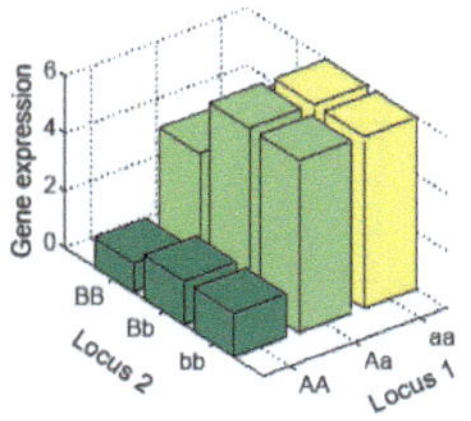

C

Interaction model
with multiplicative
effects

	BB	Bb	bb
AA	α	α	α
Aa	α	αr	αr^2
aa	α	αr^2	αr^4

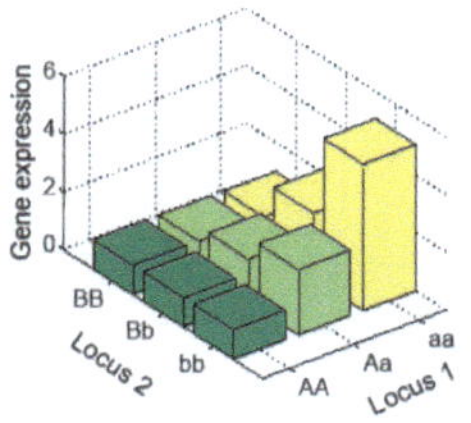

D

Interaction model
with threshold effects

	BB	Bb	bb
AA	α	α	α
Aa	α	αr	αr
aa	α	αr	αr

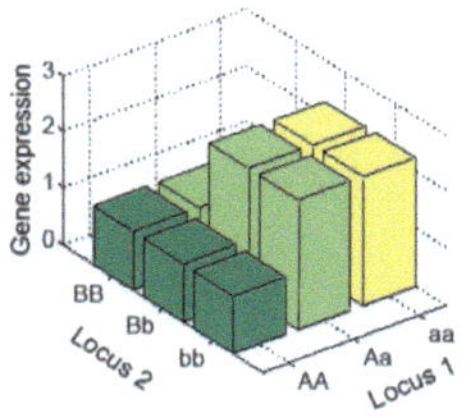

Fig. 6 Four scenarios of eQTL epistasis effects between two loci [49]. (**a**) multiplicative model within/between loci, (**b**) multiplicative model with threshold within/between loci, (**c**) interaction model with multiplicative effects, and (**d**) interaction model with threshold effects

3.4 eQTL Epistasis

eQTL epistasis is a nonlinear interaction among multiple genetic variations that play an important role in regulating gene expression. eQTL epistasis study can explain "missing heritability" that eQTL analysis does not account for [49]. Moreover, eQTL epistasis can identify causal relationships between genes in biological pathways [50–52].

Fisher's interaction model is the most prevalent mathematical model for interaction effects, where the interaction effects are modeled in a multiplicative fashion:

$$y = \beta_0 + \beta_1 x_1 + \beta_2 x_2 + \beta_{12} x_1 x_2 + \varepsilon,$$

where y is gene expressions, β_0 is an intercept, x_1 and x_2 are the numbers of minor alleles of two SNPs (in an additive model), and $\varepsilon \sim N(0, \sigma^2)$ is a residual. β_1 and β_2 are the coefficients of the main effects of each SNP, and β_{12} is the coefficient of the interaction effect of the SNPs. Fisher's model performs a significance test with the null hypothesis model that has only terms of main effects. The model detects the significant interaction between two SNPs if the interaction effect term is statistically significant. However, it lacks casual relationship between genotypic effects, and it assumes that the effect size of each SNP is equal.

The following four scenarios of interactions between two loci are introduced for eQTL epistasis [49, 53]: (1) a multiplicative model within/between loci (*Model I*), (2) a multiplicative model with threshold within/between loci (*Model II*), (3) an interaction model with multiplicative effects (*Model III*), and (4) an interaction model with threshold effects (*Model IV*). Model I proposes a multiplicative model within and between loci (Fig. 6a). This model considers the different genetic effect sizes (r_1 and r_2) of two SNPs (e.g., SNP_1 and SNP_2), and the effect sizes increase in a multiplicative fashion on the baseline effect (α) which is the background effect size when no minor alleles exist. Model II is based on Model I, but the same effect sizes are present when minor allele occurs no matter how many minor alleles are (Fig. 6b). Model III considers the equal effect size at the two loci ($r_1 = r_2 = r$). Model III is equivalent to Fisher's interaction model (Fig. 6c). Model IV is derived from Model III, but takes equal effect sizes when a minor allele is observed (Fig. 6d).

The various eQTL epistasis models are mathematically generalized by:

$$y = \beta_0 + \beta_1 x_1 + \beta_2 x_2 + \beta_{12} r_1^{\varphi(x_1)} r_2^{\varphi(x_2)} + \varepsilon,$$

where r_1 and r_2 are genetic effect sizes of SNPs (which are parameters of the model), and $\varphi(x_1)$ is the model function that represents the power of effect size in the different fashions of eQTL epistasis models. $\varphi(x_1)$ for Model I returns 2 if x_1 is *aa*, 1 if x_1 is *Aa*, otherwise 0. For instance, if two SNPs are *Aa* and *bb*, the interaction term with Model I will be $\beta_{12} r_1^1 r_2^2$, as shown in Fig. 6a. The significance of the interaction term is tested with the null hypothesis that includes main effect terms only.

When the interaction term shows its statistically significance, it tests the significance of each genetic effect size again for inferring the hierarchical relationship between the loci:

$$y = \beta_0 + \beta_1 x_1 + \beta_2 x_2 + \beta_{r_1} r_1^{\varphi(x_1)} + \beta_{r_2} r_2^{\varphi(x_2)} + \varepsilon.$$

The significant β_{r_1} or β_{r_2} represents the dominance of the genetic effect size in the interaction effect. The observation of only one significant genetic effect with an insignificant one is of interest for hierarchical relationship between the loci. Suppose that the significant epistatic effects of SNP1 and SNP2 are shown to regulate a gene G, but only SNP1 is significant for individual significant test. It can be interpreted as SNP2 $\dashv$ SNP1 $\rightarrow$ G, which means SNP1 is epistatic to SNP2. It shows that SNP1 is a downstream of SNP2 for regulating gene G.

eQTL epistasis studies can eventually construct gene regulatory networks that represent biological causal relationship between genes and provide invaluable interpretation in biological research.

4 Notes

1. An open-source whole genome association analysis tool, PLINK, (http://zzz.bwh.harvard.edu/plink) provides various analysis tools for large-scale genomic data and massive tutorials and protocols [54].
2. The multi-omics data may have relatively limited amount of available data. The multiple types of omics data should be simultaneously obtained from the same individuals. The publicly available databases include:
   (a) The Cancer Genome Atlas (TCGA): https://cancergenome.nih.gov/
   (b) cBioPortal for cancer genomics: http://www.cbioportal.org
   (c) Multi-Omics Profiling Expression Database (MOPED): https://omictools.com
   (d) PsychENCODE for psychiatric disorders: http://psychencode.org
   (e) Wellcome Trust Case Control Consortium: https://www.wtccc.org.uk

References

1. The 1000 Genomes Project Consortium (2015) A global reference for human genetic variation. Nature 526(7571):68–74
2. Rockman MV, Kruglyak L (2006) Genetics of global gene expression. Nat Rev Genet 7 (11):862–872
3. O'Connor C, Adams JU (2010) Essentials of cell biology. Nat Educ:1–100
4. Gutierrez-Arcelus M et al (2015) Tissue-specific effects of genetic and epigenetic variation on gene regulation and splicing. PLoS Genet 11(1):e1004958
5. Hoheisel JD (2006) Microarray technology: beyond transcript profiling and genotype analysis. Nat Rev Genet 7:200–210
6. Goodwin S, McPherson JD, McCombie WR (2016) Coming of age: Ten years of next-generation sequencing technologies. Nat Rev Genet
7. Sun W, Hu Y (2013) eQTL mapping using RNA-seq data. Stat Biosci 5(1):198–219
8. Kristensen VN, Lingjaerde OC, Russnes HG, Vollan HK, Frigessi A, Borresen-Dale AL (2014) Principles and methods of integrative genomic analyses in cancer. Nat Rev Cancer 14 (5):299–313
9. Zhang W, Li F, Nie L (2010) Integrating multiple 'omics' analysis for microbial biology: application and methodologies. Microbiology 156(2):287–301
10. Higdon R et al (2015) The promise of multi-omics and clinical data integration to identify and target personalized healthcare approaches in autism spectrum disorders. OMICS 19 (4):197–208
11. Rebollar EA et al (2016) Using 'omics' and integrated multi-omics approaches to guide probiotic selection to mitigate

chytridiomycosis and other emerging infectious diseases. Front Microbiol 7:68

12. Cisek K, Krochmal M, Klein J, Mischak H (2016) The application of multi-omics and systems biology to identify therapeutic targets in chronic kidney disease. Nephrol Dial Transplant 31(12):2003–2011
13. Breckpot J, Thienpont B, Gewillig M, Allegaert K, Vermeesch JR, Devriendt K (2012) Differences in copy number variation between discordant monozygotic twins as a model for exploring chromosomal mosaicism in congenital heart defects. Mol Syndromol 2 (2):81–87
14. Henrichsen CN, Chaignat E, Reymond A (2009) Copy number variants, diseases and gene expression. Hum Mol Genet 18(R1): R1–R8
15. Aure MR et al (2013) Individual and combined effects of DNA methylation and copy number alterations on miRNA expression in breast tumors. Genome Biol 14(11):R126
16. Wagner JR, Busche S, Ge B, Kwan T, Pastinen T, Blanchette M (2014) The relationship between DNA methylation, genetic and expression inter-individual variation in untransformed human fibroblasts. Genome Biol 15 (2):R37
17. Kang M, Kim DC, Liu C, Gao J (2015) Multiblock discriminant analysis for integrative genomic study. Biomed Res Int 2015:783592
18. Kim D-C, Kang M, Zhang B, Wu X, Liu C, Gao J (2014) Integration of DNA methylation, copy number variation, and gene expression for gene regulatory network inference and application to psychiatric disorders. IEEE Int Conf Bioinforma Bioeng 2014:238–242
19. Kang M, Park J, Kim DC, Biswas A, Liu C, Gao J (2017) Multi-block bipartite graph for integrative genomic analysis. IEEE/ACM Trans Comput Biol Bioinform 14:1350–1358
20. Freeman JL et al (2006) Copy number variation: new insights in genome diversity. Genome Res 16(8):949–961
21. Girirajan S, Campbell CD, Eichler EE (2011) Human copy number variation and complex genetic disease. Annu Rev Genet 45 (1):203–226
22. Gal-Yam EN et al (2008) Frequent switching of Polycomb repressive marks and DNA hypermethylation in the PC3 prostate cancer cell line. Proc Natl Acad Sci U S A 105 (35):12979–12984
23. Moore LD, Le T, Fan G (2013) DNA methylation and its basic function. Neuropsychopharmacology 38(1):23–38
24. Meissner A et al (2008) Genome-scale DNA methylation maps of pluripotent and differentiated cells. Nature 454(7205):766–770
25. Bush WS, Moore JH (2012) Chapter 11: genome-wide association studies. PLoS Comput Biol 8(12):e1002822
26. Reich DE et al (2001) Linkage disequilibrium in the human genome. Nature 411 (6834):199–204
27. Cho S, Kim H, Oh S, Kim K, Park T (2009) Elastic-net regularization approaches for genome-wide association studies of rheumatoid arthritis. BMC Proc 3(Suppl 7):S25
28. Waldmann P, Mészáros G, Gredler B, Fuerst C, Sölkner J (2013) Evaluation of the lasso and the elastic net in genome-wide association studies. Front Genet 4:270
29. Goeman JJ, Solari A (2014) Multiple hypothesis testing in genomics. Stat Med 33 (11):1946–1978
30. Cheung VG et al (2010) Polymorphic cis- and trans-regulation of human gene expression. PLoS Biol 8(9):e1000480
31. Nica AC, Dermitzakis ET (2013) Expression quantitative trait loci: present and future. Philos Trans R Soc Lond Ser B Biol Sci 368 (1620):20120362
32. Albert FW, Kruglyak L (2015) The role of regulatory variation in complex traits and disease. Nat Rev Genet 16(4):197–212
33. Michalak P (2008) Coexpression, coregulation, and cofunctionality of neighboring genes in eukaryotic genomes. Genomics 91 (3):243–248
34. Chun H, Keles S (2009) Expression quantitative trait loci mapping with multivariate sparse partial least squares regression. Genetics 182 (1):79–90
35. Lee S, Zhu J, Xing E (2010) Adaptive multitask Lasso: with application to eQTL detection. Adv Neural Inf 1:1306–1314
36. Zhang W, Zhu J, Schadt EE, Liu JS (2010) A Bayesian partition method for detecting pleiotropic and epistatic eQTL modules. PLoS Comput Biol 6(1):e1000642
37. Wittkopp PJ, Kalay G (2011) Cis-regulatory elements: molecular mechanisms and evolutionary processes underlying divergence. Nat Rev Genet 13(1):59–69
38. Huang T, Cai YD (2013) An Information-Theoretic Machine Learning Approach to Expression QTL Analysis. PLOS ONE 8(6): e67899
39. Ritchie MD, Holzinger ER, Li R, Pendergrass SA, Kim D (2015) Methods of integrating data

to uncover genotype–phenotype interactions. Nat Rev Genet 16(2):85–97

40. Wang D, Gu J (2016) Integrative clustering methods of multi-omics data for molecule-based cancer classifications. Quant Biol 4 (1):58–67
41. Zhang W et al (2013) Integrating genomic, epigenomic, and transcriptomic features reveals modular signatures underlying poor prognosis in ovarian cancer. Cell Rep 4(3):542–553
42. Zhang Z et al (2016) Molecular subtyping of serous ovarian cancer based on multi-omics data. Sci Rep 6:26001
43. Kang M, Kim DC, Liu C, Zhang B, Wu X, Gao J (2014) Multi-block and multi-task learning for integrative genomic study. In: Proceedings—IEEE 14th International Conference on Bioinformatics and Bioengineering, BIBE 2014. IEEE Computer Society, Washington, DC, pp 38–45
44. Zhang S, Liu CC, Li W, Shen H, Laird PW, Zhou XJ (2012) Discovery of multi-dimensional modules by integrative analysis of cancer genomic data. Nucleic Acids Res 40 (19):9379–9391
45. Yang Z, Michailidis G (2015) A non-negative matrix factorization method for detecting modules in heterogeneous omics multi-modal data. Bioinformatics 32(1):1–8
46. Gregory KB, Momin AA, Coombes KR, Baladandayuthapani V (2014) Latent feature decompositions for integrative analysis of multi-platform genomic data. IEEE/ACM Trans Comput Biol Bioinforma 11 (6):984–994
47. Chung R, Kang C (2019) A multi-omics data simulator for complex disease studies and its application to evaluate multi-omics data analysis methods for disease classification. GigaScience 8(5):giz045
48. Furey TS (2012) ChIP-seq and beyond: New and improved methodologies to detect and characterize protein-DNA interactions. Nat Rev Genet
49. Kang M, Zhang C, Chun HW, Ding C, Liu C, Gao J (2015) EQTL epistasis: detecting epistatic effects and inferring hierarchical relationships of genes in biological pathways. Bioinformatics 31(5):656–664
50. Aylor DL, Zeng ZB (2008) From classical genetics to quantitative genetics to systems biology: modeling epistasis. PLoS Genet 4(3): e1000029
51. Cordell HJ (2002) Epistasis: what it means, what it doesn't mean, and statistical methods to detect it in humans. Hum Mol Genet 11 (20):2463–2468
52. Phenix H et al (2011) Quantitative epistasis analysis and pathway inference from genetic interaction data. PLoS Comput Biol 7(5): e1002048
53. Marchini J, Donnelly P, Cardon LR (2005) Genome-wide strategies for detecting multiple loci that influence complex diseases. Nat Genet 37(4):413–417
54. Purcell S, Neale B, Todd-Brown K, Thomas L, Ferreira MAR, Bender D, Maller J, Sklar P, de Bakker PIW, Daly MJ, Sham PC (2007) PLINK: a toolset for whole-genome association and population-based linkage analysis. American Journal of Human Genetics, 81

Chapter 12

Sparse Partial Least Squares Methods for Joint Modular Pattern Discovery

Jinyu Chen and Shihua Zhang

Abstract

The underlying relationship between genomic factors and the response of diverse cancer drugs still remains unclear. A number of studies showed that the heterogeneous responses to anticancer treatments of patients were partly associated with their specific changes in gene expression and somatic alterations. However, how to identify the multiple-to-multiple relationships between genomic factors and drug response among pharmacogenomics data is still a challenging issue. Here, we introduce a sparse partial least squares (SPLS) framework with or without the network-regularized penalty to identify joint modular patterns demonstrated with a large-scale pairwise gene-expression and drug-response data. The identified modular patterns reveal some coordinated gene–drug associations. SPLS methods could be applied to many biological problems such as the eQTL analysis, which is designed to discover genetic variants that influence downstream gene expression level. In summary, SPLS-based methods are a set of powerful tools to uncover the associations between different types of features.

Key words Bioinformatics, Cancer genomics, Partial least squares, Integrative analysis, Network-regularized penalty

1 Introduction

The increasing amount of available high-throughput data at both or multiple levels of genomics as well as others provide us the opportunities for large-scale integrative analysis by computational methods [1–3]. A typical example is that large-scale genomic and drug response data have been generated. This situation enables us to study the underlying mechanisms of drug actions from the perspective of gene regulation. Previous studies suggest that drug molecules often interact with multiple targets, and the same mechanism of action or target is shared by more than one drug. The multiple-to-multiple relationships between drugs and their targets imply that it is valuable to discover combinatorial gene–drug patterns to gain novel insights into molecular mechanisms and examine new drug targets for therapy.

Xinghua Mindy Shi (ed.), *eQTL Analysis: Methods and Protocols*, Methods in Molecular Biology, vol. 2082,
https://doi.org/10.1007/978-1-0716-0026-9_12,

A number of large-scale drug screening projects successively start up, which provide valuable resources to reveal gene–drug associations. For example, the NCI-60 project (http://discover.nci.nih.gov/cellminer/home.do), employing an ensemble of 60 human cancer cell lines to screen over 100,000 chemical compounds and natural products [1]; The Cancer Cell Line Encyclopedia (CCLE) [2] and Genomics of Drug Sensitivity in Cancer (GDSC) [3], publishing diverse types of genomic data and pharmacological data across hundreds of cell lines. At the same time, several integrative methods have been proposed to uncover the associations between genomic predictors and drug response. For example, Barretina et al. [2] and Garnett et al. [3] both applied the elastic net method to discover genomic predictors of each drug independently; Kutalik et al. [4] developed the Ping-Pong Algorithm (PPA) to identify gene–drug co-modules using pairwise gene expression data and drug response data. Considering that gene interaction network is useful prior knowledge, which could provide valuable combinational signals to improve the module discovery accuracy, but was not yet used in these studies, we develop the sparse network-regularized partial least square (SNPLS) method recently [5]. This method integrates gene expression and drug response data across a set of cell lines as well as a gene interaction network (Fig. 1). As a result, this method enables us to identify gene–drug co-modules which are significantly related to known functions, cancers, and have coordinated gene–drug associations. These gene–drug patterns will provide us insights into potential drug targets and drug combinations for cancer therapy.

More importantly, such sparse partial least squares (SPLS) are expected to be adapted to diverse biological problems to discover joint modular patterns [6]. For example, Li and Zhang have introduced a sparse multi-block partial least squares (sMBPLS) regression method to identify multidimensional regulatory modules from multidimensional genomic data such as copy number variation, DNA methylation, microRNA expression, and gene expression data [7]. One another applicable topic is about the so-called GWAS as well as eQTL. With the sequencing technological advances, it is becoming an essential topic to explore the disease or some organismal phenotypes from the perspective of genome variants. Thus, a large number of genome-wide association studies (GWAS) [8–10] spring up to identify genetic variants that contribute to some phenotypes. Gene expression levels can be viewed as quantitative phenotypes. Exploring the relationships between genome variants and gene expression levels will provide more direct insight into the molecular mechanisms of complex diseases. The analysis of identifying genetic variants that explain variation in gene expression levels has spawned a new field named expression quantitative trait loci (eQTL) analysis. Standard eQTL analysis involves a direct association test between genetic variation markers and gene

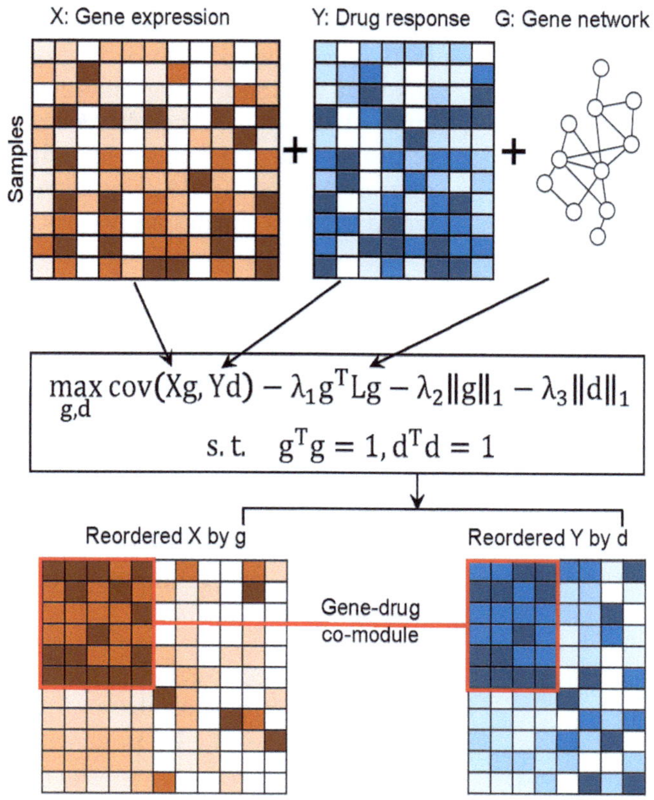

Fig. 1 Overview of the sparse network-regularized partial least squares method (SNPLS) for identifying gene–drug co-modules. A co-module is a subset of genes and drugs exhibiting similar profiles across a subset of samples determined by the weight variables *g* and *d* of SNPLS applied in pairwise gene expression data *X* and drug response data *Y*. A gene interaction network *G* is incorporated to enhance the modular characteristics

expression levels across the same set of hundreds of individuals. In the univariate methods, only one marker is tested at a time whether it is correlated with the expression of the target gene. The relationship is measured by a log-odds (LOD) score [11, 12]. However, such methods ignore the combinatorial effects of sets of single-nucleotide polymorphism (SNP) markers. Thus the class of multivariate methods are proposed to identify pairs or more generally sets of markers which together explain the gene expression variations. For example, Broman and Speed [13] applied the multivariate linear regression model to eQTL problem; Bureau et al. [14] and Lee et al. [15] both used random forests to the problem of identifying SNPs to explain phenotypes, which consider the nonlinear interactions. Chun and Keles [16] employed a sparse partial least squares regression method to select SNP markers associated with each cluster of genes, which simultaneously perform variable selection and dimension reduction for eQTL mapping problem.

In this chapter, we introduce a sparse partial least squares (SPLS) framework with or without the network-regularized penalty term to identify joint modular patterns and demonstrate it with an application in large-scale pairwise gene-expression and drug-response data. Further, we introduce a package called Matrix Integration Analysis (MIA), implementing and extending four published methods including two partial least squares (PLS) ones.

2 Materials

2.1 Gene Expression and Drug Response Data

We downloaded a large-scale pharmacogenomic dataset including pairwise gene expression data and drug response data from the GDSC website (http://www.cancerrxgene.org/) [3]. After removing several drugs and samples that only have a limited number of values across samples and drugs, respectively, we obtained a gene expression dataset of 13,321 genes and a drug response dataset of 98 drugs across 641 samples, which are represented in two matrices $X \in \mathbb{R}^{p \times n}$ and $Y \in \mathbb{R}^{p \times m}$, respectively.

2.2 Gene Interaction Network

We downloaded a gene or protein interaction network from the PathwayCommons database (http://www.pathwaycommons.org) [17]. It consists of 14,355 genes or proteins and 507,757 interactions. We filtered the genes which are absent from the gene expression data. For any gene that is in the input gene expression data X but not in this network, we added it to the network as an isolated node. Finally, we obtained a gene–gene interaction network with 13,321 genes and 262,462 interactions, which are denoted as a graph G. The adjacency matrix of this graph is denoted as A, where $A_{ij} = 1$, if gene i and j are linked in the network and $A_{ij} = 0$, otherwise.

3 Methods

Partial least squares method (PLS) is a multivariate regression method used to find the fundamental relations between an input matrix $X \in \mathbb{R}^{p \times n}$ and a response matrix $Y \in \mathbb{R}^{p \times m}$, both of which have the same rows (samples). Comparing to the classical linear regression, it works well for the data with small sample sizes and a large number of variables ($p < n$). PLS decomposes matrix X and Y both with zero-mean variables into the form: $X = \mathrm{UP}^T + E$, $Y = VQ^T + F$, where U, V are ($p \times k$) matrices of k latent components which describe the original data matrices in a lower space, and U and V can be constructed as a linear transformation of X, Y, respectively; P, Q represent loading matrices to measure the relationships between original variables and latent ones; E, F are the residual matrices. PLS prefers to maximize the covariance between X and Y by means of latent components U and V.

3.1 The Standard PLS Algorithm

The objective function is

$$\max_{g,\,d} \operatorname{cov}(Xg,\ Yd)$$

$$\text{s.t. } g^T g = 1,\ \ d^T d = 1. \tag{1}$$

Here, let's denote $u = Xg$, $v = Yd$ as the latent variables which are the linear combination of n and m variables corresponding to X and Y, respectively. g and d are also named as weight vectors. This objective function indicates that the similarity between small blocks of X and Y is measured by the covariance of the two latent variables u and v. Note that, since $[\text{cov(Xg, Yd)}]^2 = \text{var}\,(Xg)[\text{corr}(Xg, Yd)]^2\,\text{var}\,(Yd)$, thus, the weight vectors g and d identified by PLS simultaneously take into account the requirements of maximal correlation between X and Y like Canonical Correlation Analysis (CCA) and to explain as much variance as possible in both X- and Y-space like Principal Components Analysis (PCA). Finally, based on the absolute values of the optimal solutions of g and d, we could discover the corresponding blocks in X and Y which have similar or coherent patterns.

3.2 The Sparse PLS Algorithm

Since standard PLS method does not perform variable selection and is likely to result in poor interpretation, Chun and Keles [18] suggested to impose a sparsity penalty to the weight variables g and d and developed a sparse PLS regression (SPLS) method, which is also extended for multiple genomics data analysis recently [7], named as sMBPLS. Li et al. applied sMBPLS regression method to multiple genomic datasets, including copy number variation, DNA methylation, microRNA expression, and gene expression data, in order to identify a multidimensional regulatory module jointly contributing to a local "gene expression factory". sMBPLS model is designed as follows:

$$\max_{g,\,d} cov\left(\sum_{i=1}^{n} b_i X_i g_i,\ Yd\right) - \sum_{i=1}^{n} \lambda_i \|g_i\|_1 - \mu\|d\|_1$$

$$\text{s.t. } \sum_i b_i^{\,2} = 1,\quad g_i^{\,T} g_i = 1,\quad d^T d = 1,\quad \text{for } i = 1, \ldots, n.$$

where g_i is the loading vector for the ith type of input data block X_i, and d for Y. b_i is the weight for data block X_i when identifying the relationships between multiple input data blocks X_i and response data Y. The SPLS produces sparse g_i and d, which can be used for selecting effective variables with better biological interpretation. When $n = 1$, this model reduces to the basic sparse PLS framework.

3.3 The Sparse Network-Regularized PLS (SNPLS) Framework

In the issue that uncovers the associations between drugs and genes, prior knowledge on gene interactions is very useful and valuable to decipher the modular patterns among genes. Network-based penalty has been adopted for many different applications. For example, Li and Li [19, 20] and Liu et al. [21] developed a network-constrained regularization procedure to realize variable selection and regression analysis for genomic data. Zhang et al. [22] utilized predicted miRNA–gene interactions and gene–gene interactions to define network-regularized constraints for discovering the miRNA–gene regulatory modules. These network-based penalty functions all aim at enforcing the tightly connected nodes (genes) in the network to have more similar coefficients. Inspired by this technique, we introduced a sparse network-regularized partial least square (SNPLS) model to achieve our goal. Specifically, it can be formulated as follows:

$$\max_{g,d} \operatorname{cov}(Xg, Yd) - \lambda_1 g^T Lg - \lambda_2 ||g||_1 - \lambda_3 ||d||_1$$

$$\text{s.t. } g^T g = 1, \ d^T d = 1. \tag{2}$$

where $\operatorname{cov}(u,v)$ is the covariance of u and v ($u, v \in \mathbb{R}^p$), which approximates to $\frac{1}{p} u^T v$, if $\sum_{i=1}^{p} u_i = \sum_{i=1}^{p} v_i = 0$, and L is the symmetric normalized Laplacian matrix defined as

$$L = D^{-\frac{1}{2}}(D - A)D^{-\frac{1}{2}} = I - D^{-\frac{1}{2}}AD^{-\frac{1}{2}},$$

where D is the degree matrix of graph G, that is, $d_{ii} = \sum_{j=1}^{n} a_{ij}$, and $d_{ij} = 0$, for $i \neq j$. The tuning parameters λ_1, λ_2, λ_3 control the amount of regularization for smoothness and sparsity. When $\lambda_1 = 0$, the model reduces to the SPLS.

If the matrices X and Y are normalized such that each column of X and Y is centered, the problem defined in Eq. (2) is equivalent to

$$\max_{g,d} \frac{1}{p} g^T X^T Y d - \lambda_1 \sum_{1 \le i < j \le n} a_{ij} \left(\frac{g_i}{\sqrt{l_i}} - \frac{g_j}{\sqrt{l_j}} \right)^2 - \lambda_2 ||g||_1 - \lambda_3 ||d||_1$$

$$\text{s.t. } g^T g = 1, \ d^T d = 1.$$

The objective function consists of four key terms. The first one describes the covariance between the hidden components based on the gene expression data X and drugs response data Y. The second one captures the key prior knowledge which makes the connected genes in the network likely to be placed in the same co-modules. The last two ones enforce the sparsity of the variables g and d such that the results will have better biological interpretation.

3.4 Algorithms

In the literature, standard PLS model can be solved by nonlinear iterative partial least squares (NIPALS) algorithm [6]. In terms of SNPLS model, the network-regularized term and sparse penalties make it difficult to be solved. Here, we developed a coordinate descent algorithm to find local maximum of this problem by updating variables g and d alternately. To speed up the convergence of this algorithm, we employed the solution of standard PLS as the initial solution of current algorithm.

Step 1: Initialize g with the solution of Eq. (1) and $u = Xg$.

Step 2: Update d and g alternately.

1. Fix variable g and update variable d with

$$d \leftarrow sign\left(\frac{1}{p}\Upsilon^T u\right)\left(\left|\frac{1}{p}\Upsilon^T u\right| - \lambda_3\right)_+, \quad norm\ d.$$

$$v = \Upsilon d$$

where

$$(x)_+ = \begin{cases} x, & \text{if } x > 0, \\ 0, & \text{otherwise.} \end{cases}$$

2. Fix variable d and update variable g with

$$g \leftarrow \frac{sign(z)(|z| - \lambda_2)_+}{2(\lambda_1 + \delta)}, \quad \text{for } j = 1, 2, \ldots, n;\ norm\ g.$$

$$u = Xg.$$

where $z = t_j + 2\lambda_1 \sum_{i=1}^{n} \frac{a_{ij}g_i}{\sqrt{l_i l_j}}$, and t_j is the jth element of vector $t = \frac{1}{p}(X^T \Upsilon d) = \frac{1}{p}(X^T v)$. δ is a positive parameter for the constraint $g^T g = 1$.

Step 3: Repeat Step 2 until convergence of u.

Obviously, the computational complexity of one SNPLS iteration is $O(pm + pn + n^2)$. We implemented it in MATLAB R2013a as a user-friendly package (http://page.amss.ac.cn/shihua.zhang/).

3.5 Determining Co-modules

The weight vectors g and d produced by the above algorithm will guide us to identify gene–drug co-modules. The main idea is to select the gene and drug variables with relatively large absolute values of weight variables g and d as the members of gene–drug co-modules. Specifically, we calculated the z-scores of g and d in the following way:

$$z_i = \frac{||x_i| - \overline{x}|}{S_x} \tag{3}$$

where $\overline{x} = \frac{1}{n}\sum_i |x_i|$, $S_x^2 = \frac{1}{n-1}\sum_i (|x_i| - \overline{x})^2$.

Based on this transformation, we obtained two vectors g^* and d^* and determined the co-module members if $g^*(i)$ (or $d^*(j)$) was larger than the given threshold T. Meanwhile, we updated g and d by setting the values of g_i and d_j which were not selected as the members of a co-module be zeros. Moreover, we prefered to identify the gene–drug co-modules across certain subset of samples. To achieve this goal, we considered the latent vectors $u = Xg$ and $v = Yd$ and normalized u and v, such that $u^* = \frac{u}{||u||_2}$ and $v^* = \frac{v}{||v||_2}$. We applied formula (3) to the vector $(u^* + v^*)$, chose samples whose scores were larger than a given threshold, and updated u and v as what we did to g and d.

We obtained the first gene–drug co-module after running the SNPLS algorithm. Next, we subtracted the signal of current co-module from the input data as follows:

$$X := X - up^{T}, \quad p = \frac{X^{T}u}{u^{T}u}$$

$$Y := Y - vq^{T}, \quad q = \frac{Y^{T}v}{v^{T}v}$$

Then, we could continue to apply SNPLS algorithm to the updated input data X and Y to identify the next gene–drug co-module.

3.6 Overlap Significance Test Between Co-modules

Each module consists of three feature sets including genes, drugs, and cell lines. We applied the right-tailed hypergeometric test to test the overlap significance of each feature set of any two co-modules. If more than one of these three tests are significant, that is, the q-value with multiple testing correction is smaller than 0.05, we considered these two co-modules were significantly overlapped.

3.7 Enrichment Test of Gene Interactions in Co-modules

We tested whether the genes in each co-module were tightly connected in the gene–gene interaction network. If the network used in the SNPLS method has n genes, p edges and the ith co-module contains n_i genes and k edges in the network, p-value of this test is $\sum_{x \geq k} \frac{C_p^x C_{N-p}^{M-x}}{C_N^M}$, where $N = C_n{}^2$, $M = C_{n_i}{}^2$. After multiple testing correction, we obtained q-value for the gene set in each co-module. We considered the gene set was significantly tightly connected with $q\text{-value} < 0.05$.

3.8 Functional Analysis of Identified Gene–Drug Co-modules

To evaluate the biological relevance among the three types of components—genes, drugs, and cell lines, we firstly performed systematic enrichment analysis in terms of GO biological process and KEGG pathway for gene components by DAVID website [23]; then, by searching for some databases (e.g., DrugBank [24], KEGG [25], and PharmGKB [26]) and published studies, we summarized the drug targets, drug effector pathways, or biological processes for drug components; for cancer cell lines, we tested whether they tended to belong to the same or similar tumor types; finally, we investigated their connections, for example, in one co-module, whether the gene set contains some drug targets; whether the drugs target the gene-enriched biological processes or KEGG pathway; or whether the drugs just have effects on the tumor types enriched in the selected cell lines.

3.9 Construction of Co-module Networks with Three Levels

To demonstrate the relationships between different levels, we could construct three-level network for each co-module. Firstly, we used the GeneMANIA prediction server [27] to connect genes, including three types of high confidential interactions: physical interactions, pathway interactions (two genes are linked if they participate in the same reaction within a pathway), and shared protein domains interactions (two genes are linked if they have the same protein domain). Then, we computed the Pearson correlation coefficients between genes and drugs using the pharmacogenomic data X and Y and selected the top 10% as gene–drug interactions. Next, by manually searching literature, we linked the drugs and tumor types if a drug can be used to treat a given cancer. Finally, we removed the genes which have no links with other genes and keeped the maximal connected component.

4 Results

We applied SNPLS to a large-scale pharmacogenomic dataset derived from the Genomics of Drug Sensitivity in Cancer (GDSC) [3] and obtained 20 gene–drug co-modules. The 20 gene–drug co-modules cover about 30 cell lines, 137 genes, and 2 drugs on average. We found that each of the three components occurred in about one to three co-modules, indicating that the 20 co-modules are different with each other. Furthermore, by means of overlap significance test between any two co-modules, it demonstrates that only one pair of co-modules has significant overlap (FDR < 0.05). Thus, almost all of the 20 co-modules are distinct. The co-module member genes and drugs also exhibit highly similar patterns across the same subset of samples (e.g., co-module 1 in Fig. 2a, b).

4.1 Co-modules Reveal Distinct Biological Relevance

For the gene modules, 12 (60%) and 11 (55%), respectively, have at least one significant GO biological process and KEGG pathway (Benjamini-corrected p value < 0.05). In total, these modules are enriched in 193 distinct GO biological processes and 29 KEGG pathways. Among them, the most frequent biological processes are biological adhesion, chromosome organization, cell cycle, and mitosis. The most frequently enriched KEGG pathways are focal adhesion and cell cycle.

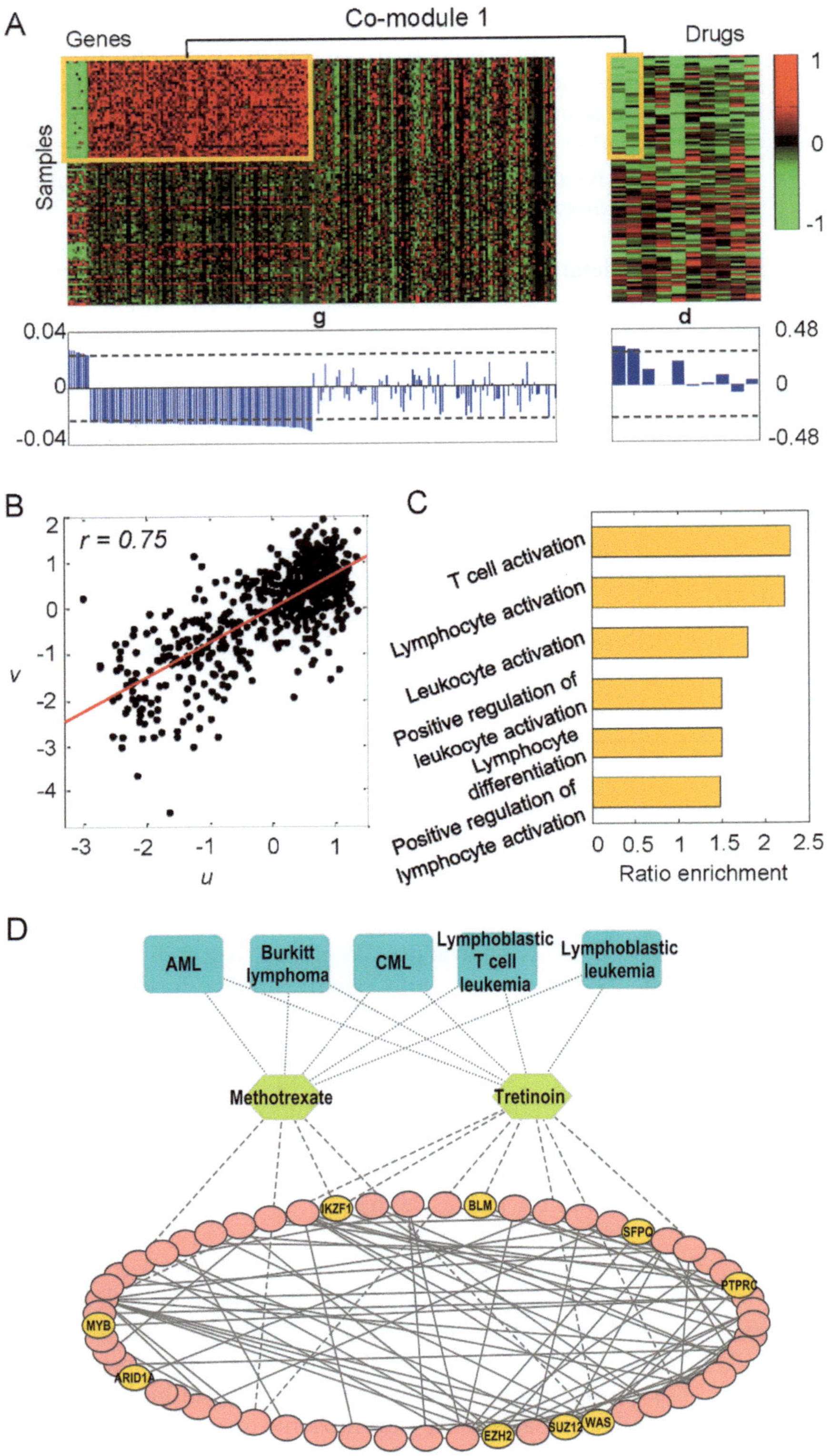
A
Co-module 1
Genes
Drugs
Samples
1
0
-1
g
d
0.04
0
-0.04
0.48
0
-0.48
B
r = 0.75
v
u
C
T cell activation
Lymphocyte activation
Leukocyte activation
Positive regulation of leukocyte activation
Lymphocyte differentiation
Positive regulation of lymphocyte activation
Ratio enrichment
D
AML
Burkitt lymphoma
CML
Lymphoblastic T cell leukemia
Lymphoblastic leukemia
Methotrexate
Tretinoin
IKZF1
BLM
SFPQ
PTPRC
MYB
ARID1A
EZH2
SUZ12
WAS

For the drug co-modules, 12 of the 20 co-modules include more than one drug. Among 6 (50%) of these 12 drug modules, their drug members share the same targets or effector pathways. For example, the three drugs (CI-1040, PLX4720 and SB590885) in co-module 11 all target *ERK* signaling pathway.

Moreover, for samples in the 20 co-modules, 13 (65%) modules are enriched in certain tumor types (FDR < 0.05, hypergeometric test). For example, the co-module 1 is enriched in several types of blood diseases including lymphoblastic T cell leukemia, lymphoblastic leukemia, acute myelogenous leukemia (AML), Burkitt lymphoma, chronic myelogenous leukemia (CML), and so on.

4.2 Co-modules Reveal Significant Drug–Gene Connections

We found that the co-modules reveal significant drug–gene connections from different angles [5]. Here, we took co-module 1 as an example. The co-module 1 consists of 104 member genes with a significant number (16) of genes in the human cancer genes census (FDR $= 2.1456 \times 10^{-6}$ hypergeometric test), and a significant number (6) of genes (*BCOR, BLM, IKZF1, PTPRC, 6-Sep, SFRS2*) relating to leukemia [28]. Moreover, 8 of 12 enriched GO biological processes are about leukocyte, lymphocyte, or T cell, which are all closely related to leukemia (Fig. 2c). Surprisingly, the sample-enriched tumor types exactly refer to this kind of disease, including lymphoblastic T cell leukemia, lymphoblastic leukemia, AML, Burkitt lymphoma, and CML, indicating the distinct biological relevance of the identified co-modules. On the other hand, its two member drugs both have effects on transcription, which is a key part of cell activation. This is consistent with the enriched biological functions: cell activation, translational elongation, and ribosome pathway which plays a leading role during transcription. Furthermore, the three-level network for co-module 1 (Fig. 2d) demonstrates their close connections among genes, drugs, and sample-enriched tumor types.

4.3 Comparison with Other Methods

In order to demonstrate the effectiveness of SNPLS, we also applied SPLS [18] method without the network information to the same pharmacogenomic data and identified 20 co-modules. We

Fig. 2 Illustration of co-module 1. (**a**) Heat map of co-module 1 consisting of 104 genes and 2 drugs across 42 samples (yellow boxes). We extended the heat map to cover more variables by randomly selecting 104 genes, 8 drugs, and 58 samples for contrasting. We reordered the genes, drugs, as well as samples in this co-module circled in yellow lines by the descending order of the weight variables *g* and *d* as shown with bar plots below the heat map. The horizontal lines over the bar plots indicate the thresholds used for selecting the co-module genes and drugs. (**b**) The scatter plot for normalized latent variables *u* and *v* of co-module 1 with Pearson correlation coefficient $r = 0.75$ indicating that they are highly correlated. (**c**) Top biological terms enriched by the genes of co-module 1. The ratio enrichment indicates the functional significance of a gene module with $-\log(p\text{-value})$ (Benjamini-corrected p-value). (**d**) A network model of three levels of genes, drugs, and tumor types in co-module 1 analyzed the interaction enrichment in each co-module of SPLS and SNPLS based on the gene–gene interaction network. 14 (70%) co-modules of SNPLS are enriched with gene interactions (FDR < 0.05), whereas only 11 (55%) co-modules of SPLS are enriched, indicating the strong biological relevance of co-modules of SNPLS than those of SPLS. Actually, 14 co-modules of SNPLS are enriched in at least one GO biological process or KEGG pathway (Benjamini-corrected q-value < 0.05) and only 11 co-modules of SPLS are. We also found that the enriched biological processes of SNPLS have more significant p-value than those of SPLS, suggesting that SNPLS indeed have improvement in identifying more biologically relevant genes [5].

5 Conclusion

Deciphering the multiple-to-multiple relationships between drugs and their targets is crucial for studying the mechanisms of drug actions and developing effective treatment for patients. Meanwhile, the dramatic accumulation of large-scale genomic data and drug response data from the same cell lines provides us the unprecedented opportunities to identify gene–drug joint modular patterns to decode these relationships from the perspective of gene regulation. Chen and Zhang have demonstrated the SNPLS method to integrate these two data as well as a gene interaction network to identify gene–drug co-modules [5]. Compared to SPLS, SNPLS employs the network structure as prior knowledge such that genes in each co-module tend to be closely connected in the network, which makes such a co-module more biologically interpretable.

More importantly, SNPLS could also be applied to many other biological problems that explore the associations between two or multiple types of features, such as the eQTL analysis. By incorporating the gene interaction network structure into the analysis, the identified gene modules will be more interpretable. Besides, prior relationships between SNPs could be added into the SNPLS model in the similar way.

In addition, for the convenience of modular and integrative analysis, we developed a MATLAB package, Matrix Integration Analysis (MIA), implementing and extending four published methods, designed based on two classical techniques, non-negative matrix factorization (NMF) and partial least squares (PLS) [29]. MIA is a flexible tool which could handle a wide range of biological problems and data types. Besides, we also provided an executable version for users without the MATLAB license. It is available at http://page.amss.ac.cn/shihua.zhang/software.html. We expect this tool including the PLS-based ones will become a powerful tool in diverse biological applications.

Acknowledgment

This work has been supported by the National Natural Science Foundation of China, No. 61379092, 61422309, 61621003, and 11131009, the Strategic Priority Research Program of the Chinese Academy of Sciences (CAS) (XDB13040600), the Outstanding Young Scientist Program of CAS, CAS Frontier Science Research Key Project for Top Young Scientist (No. QYZDB-SSW-SYS008), and the Key Laboratory of Random Complex Structures and Data Science, CAS (No. 2008DP173182).

References

1. Shoemaker RH (2006) The NCI60 human tumour cell line anticancer drug screen. Nat Rev Cancer 6(10):813–823. https://doi.org/10.1038/nrc1951
2. Barretina J, Caponigro G, Stransky N, Venkatesan K, Margolin AA, Kim S, Wilson CJ, Lehar J, Kryukov GV, Sonkin D, Reddy A, Liu M, Murray L, Berger MF, Monahan JE, Morais P, Meltzer J, Korejwa A, Jane-Valbuena J, Mapa FA, Thibault J, Bric-Furlong E, Raman P, Shipway A, Engels IH, Cheng J, Yu GK, Yu JJ, Aspesi P, de Silva M, Jagtap K, Jones MD, Wang L, Hatton C, Palescandolo E, Gupta S, Mahan S, Sougnez C, Onofrio RC, Liefeld T, MacConaill L, Winckler W, Reich M, Li NX, Mesirov JP, Gabriel SB, Getz G, Ardlie K, Chan V, Myer VE, Weber BL, Porter J, Warmuth M, Finan P, Harris JL, Meyerson M, Golub TR, Morrissey MP, Sellers WR, Schlegel R, Garraway LA (2012) The Cancer Cell Line Encyclopedia enables predictive modelling of anticancer drug sensitivity. Nature 483(7391):603--609. https://doi.org/10.1038/nature11003
3. Garnett MJ, Edelman EJ, Heidorn SJ, Greenman CD, Dastur A, Lau KW, Greninger P, Thompson IR, Luo X, Soares J, Liu Q, Iorio F, Surdez D, Chen L, Milano RJ, Bignell GR, Tam AT, Davies H, Stevenson JA, Barthorpe S, Lutz SR, Kogera F, Lawrence K, McLaren-Douglas A, Mitropoulos X, Mironenko T, Thi H, Richardson L, Zhou W, Jewitt F, Zhang T, O'Brien P, Boisvert JL, Price S, Hur W, Yang W, Deng X, Butler A, Choi HG, Chang JW, Baselga J, Stamenkovic I, Engelman JA, Sharma SV, Delattre O, Saez-Rodriguez J, Gray NS, Settleman J, Futreal PA, Haber DA, Stratton MR, Ramaswamy S, McDermott U, Benes CH (2012) Systematic identification of genomic markers of drug sensitivity in cancer cells. Nature 483 (7391):570–575. https://doi.org/10.1038/nature11005
4. Kutalik Z, Beckmann JS, Bergmann S (2008) A modular approach for integrative analysis of large-scale gene-expression and drug-response data. Nat Biotechnol 26(5):531–539. https://doi.org/10.1038/nbt1397
5. Chen J, Zhang S (2016) Integrative analysis for identifying joint modular patterns of gene-expression and drug-response data. Bioinformatics 32(11):1724–1732. https://doi.org/10.1093/bioinformatics/btw059
6. Rosipal R, Kramer N (2006) Overview and recent advances in partial least squares. In: Saunders C., Grobelnik M, Gunn S, Shawe-Taylor J (eds) Subspace, latent structure and feature selection. SLSFS 2005. Lecture notes in computer science, vol 3940. Springer, Berlin, Heidelberg. https://doi.org/10.1007/11752790_2
7. Li WY, Zhang SH, Liu CC, Zhou XJ (2012) Identifying multi-layer gene regulatory modules from multi-dimensional genomic data. Bioinformatics 28(19):2458–2466. https://doi.org/10.1093/bioinformatics/bts476
8. Albert FW, Kruglyak L (2015) The role of regulatory variation in complex traits and disease. Nat Rev Genet 16(4):197–212. https://doi.org/10.1038/nrg3891
9. Gilad Y, Rifkin SA, Pritchard JK (2008) Revealing the architecture of gene regulation: the promise of eQTL studies. Trends Genet 24 (8):408–415. https://doi.org/10.1016/j.tig.2008.06.001
10. Michaelson JJ, Loguercio S, Beyer A (2009) Detection and interpretation of expression quantitative trait loci (eQTL). Methods 48 (3):265–276. https://doi.org/10.1016/j.ymeth.2009.03.004
11. Lander ES, Botstein D (1989) Mapping mendelian factors underlying quantitative traits

using RFLP linkage maps. Genetics 121 (1):185–199. http://www.genetics.org/content/121/1/185.abstract

12. Haley CS, Knott SA (1992) A simple regression method for mapping quantitative trait loci in line crosses using flanking markers. Heredity 69:315–324. https://doi.org/10.1038/hdy.1992.131
13. Broman KW, Speed TP (2002) A model selection approach for the identification of quantitative trait loci in experimental crosses. J Roy Stat Soc B 64:641–656. https://doi.org/10.1111/1467-9868.00354
14. Bureau A, Dupuis J, Falls K, Lunetta KL, Hayward B, Keith TP, Van Eerdewegh P (2005) Identifying SNPs predictive of phenotype using random forests. Genet Epidemiol 28 (2):171–182. https://doi.org/10.1002/gepi.20041
15. Lee SSF, Sun L, Kustra R, Bull SB (2008) EM-random forest and new measures of variable importance for multi-locus quantitative trait linkage analysis. Bioinformatics 24 (14):1603–1610. https://doi.org/10.1093/bioinformatics/btn239
16. Chun H, Keles S (2009) Expression quantitative trait loci mapping with multivariate sparse partial least squares regression. Genetics 182 (1):79–90. https://doi.org/10.1534/genetics.109.100362
17. Cerami EG, Gross BE, Demir E, Rodchenkov I, Babur O, Anwar N, Schultz N, Bader GD, Sander C (2011) Pathway Commons, a web resource for biological pathway data. Nucleic Acids Res 39: D685–D690. https://doi.org/10.1093/nar/gkq1039
18. Chun H, Keles S (2010) Sparse partial least squares regression for simultaneous dimension reduction and variable selection. J Roy Stat Soc B 72:3–25. https://doi.org/10.1111/j.1467-9868.2009.00723.x
19. Li CY, Li HZ (2008) Network-constrained regularization and variable selection for analysis of genomic data. Bioinformatics 24 (9):1175–1182. https://doi.org/10.1093/bioinformatics/btn081
20. Li C, Li H (2010) Variable selection and regression analysis for graph-structured covariates with an application to genomics. Ann Appl Stat 4(3):1498–1516. https://doi.org/10.1214/10-AOAS332
21. Liu J, Huang J, Ma S (2013) Incorporating network structure in integrative analysis of cancer prognosis data. Genet Epidemiol 37 (2):173–183. https://doi.org/10.1002/gepi.21697
22. Zhang SH, Li QJ, Liu J, Zhou XJ (2011) A novel computational framework for simultaneous integration of multiple types of genomic data to identify microRNA-gene regulatory modules. Bioinformatics 27(13):I401–I409. https://doi.org/10.1093/bioinformatics/btr206
23. Huang DW, Sherman BT, Lempicki RA (2009) Systematic and integrative analysis of large gene lists using DAVID bioinformatics resources. Nat Protoc 4(1):44–57. https://doi.org/10.1038/nprot.2008.211
24. Law V, Knox C, Djoumbou Y, Jewison T, Guo AC, Liu YF, Maciejewski A, Arndt D, Wilson M, Neveu V, Tang A, Gabriel G, Ly C, Adamjee S, Dame ZT, Han BS, Zhou Y, Wishart DS (2014) DrugBank 4.0: shedding new light on drug metabolism. Nucleic Acids Res 42(D1):D1091–D1097. https://doi.org/10.1093/nar/gkt1068
25. Kanehisa M, Goto S (2000) KEGG: Kyoto encyclopedia of genes and genomes. Nucleic Acids Res 28(1):27–30. https://doi.org/10.1093/nar/28.1.27
26. Whirl-Carrillo M, McDonagh EM, Hebert JM, Gong L, Sangkuhl K, Thorn CF, Altman RB, Klein TE (2012) Pharmacogenomics knowledge for personalized medicine. Clin Pharmacol Ther 92(4):414–417. https://doi.org/10.1038/clpt.2012.96
27. Warde-Farley D, Donaldson SL, Comes O, Zuberi K, Badrawi R, Chao P, Franz M, Grouios C, Kazi F, Lopes CT, Maitland A, Mostafavi S, Montojo J, Shao Q, Wright G, Bader GD, Morris Q (2010) The GeneMANIA prediction server: biological network integration for gene prioritization and predicting gene function. Nucleic Acids Res 38:W214–W220. https://doi.org/10.1093/nar/gkq537
28. Futreal PA, Coin L, Marshall M, Down T, Hubbard T, Wooster R, Rahman N, Stratton MR (2004) A census of human cancer genes. Nat Rev Cancer 4(3):177–183. https://doi.org/10.1038/nrc1299
29. Chen J, Zhang S (2018) Matrix integrative analysis (MIA) of multiple genomic data for modular patterns. Front Genet 9:194. https://doi.org/10.3389/fgene.2018.00194

Part II

Applications

Chapter 13

Expression Quantitative Trait Loci (eQTL) Analysis in Cancer

Yaoming Liu, Youqiong Ye, Jing Gong, and Leng Han

Abstract

Expression quantitative trait loci (eQTL) analysis links variations in gene expression levels to genotypes. Analyzing both cis- and trans-eQTLs from tumor samples can provide an intermediate phenotype between genetic variation and complex traits to better understand how risk alleles contribute to tumorigenesis and development. Here we describe a detailed workflow for identifying eQTLs in cancer using existing packages and software. The key package is Matrix eQTL, which requires input data of genotypes, genes expression, and covariates. This pipeline can be easily applied in a related research field.

Key words Expression quantitative trait loci, Cancer, Matrix eQTL, Genotype, Gene expression, Covariates

1 Introduction

It has been shown unambiguously that genetic variation is a determinant of gene expression [1–4]. Single nucleotide polymorphism (SNP) is the most common type of human genetic variation [5]. Most risk SNPs found in genome-wide association studies are located in noncoding regions [6], which indicates that they function via transcriptional regulatory connections to genes. Expression quantitative trait loci (eQTLs) are genetic variants that are associated with gene transcription levels. eQTLs provide an intermediate phenotype between genetic variation and complex traits to better understand how risk alleles contribute to disease [7–10].

Most eQTL data are currently derived from cell lines and normal human tissues [11–13]. However, previous studies showed that the majority of eQTLs identified in cancer samples are cancer specific compared with normal samples [14, 15]. In addition, many eQTL studies focus only on cis-eQTLs or a small subset of trans-eQTLs [16, 17] even though SNPs linked to trans-eQTL hotspots are thought to serve important regulatory roles [18]. Therefore, it is necessary to analyze both cis- and trans-eQTLs from tumor

Xinghua Mindy Shi (ed.), *eQTL Analysis: Methods and Protocols*, Methods in Molecular Biology, vol. 2082,
https://doi.org/10.1007/978-1-0716-0026-9_13,

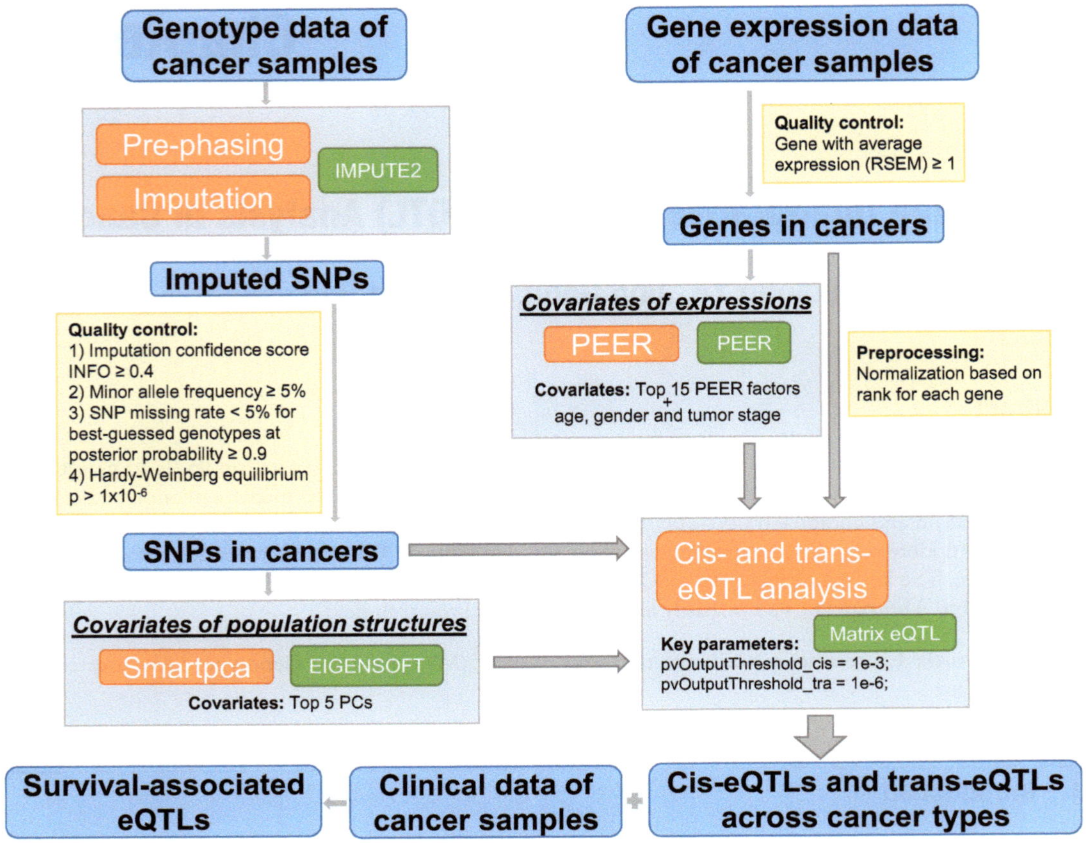

Fig. 1 The workflow of eQTL analysis in cancer

samples to further understand their functional effects in tumorigenesis.

We built a database of both cis- and trans-eQTLs in multiple cancer types using data from The Cancer Genome Atlas (TCGA) [19]. Here we describe the detailed workflow for identifying eQTLs in cancer using existing packages and software. The main process within the pipeline is summarized in Fig. 1. The key package is Matrix eQTL [20], which requires data of genotypes, genes expression, and covariates. IMPUTE2 [21] is used to impute SNP data. Smartpca in EIGENSOFT [22] is used to generate covariates from genotype data to eliminate population effects, and PEER [23] is used to find covariates from gene expression data to remove batch effects and clinical confounders. This pipeline can be easily applied in a related research field.

2 Materials

2.1 Software

1. Install IMPUTE2 [21] (https://mathgen.stats.ox.ac.uk/impute/impute_v2.html#download), which can be implemented on machines with Linux (x86_64) and Mac OSX Intel.To unpack the files on a Linux computer, use a command like this:

```
tar -zxvf  impute_v2.3.2_x86_64_static.tgz
```

2. Install EIGENSOFT [22], for which the source code, documentation and executables on a Linux platform are available here (https://www.hsph.harvard.edu/alkes-price/software/).

```
tar -zxvf  EIG-7.2.1.tar.gz
```

2.2 Package

1. PEER [23] (https://github.com/PMBio/peer/wiki) offers packages in R and python.

   For instance, install the PEER package for R on Linux by the following command:

```
R CMD INSTALL R_peer_source_1.3.tgz
```

2. Matrix eQTL [20] (http://www.bios.unc.edu/research/genomic_software/Matrix_eQTL/) is available for MATLAB and R implementations.

```
install.packages("MatrixEQTL")
```

3 Method

3.1 Genotype Data Collection, Imputation, and Processing

Genotype data of different cancer types from TCGA can be obtained via dbGaP (Study Accession of TCGA: phs000178.v10.p8. https://www.ncbi.nlm.nih.gov/projects/gap/cgi-bin/study.cgi?study_id=phs000178.v10.p8), which contains SNP data obtained using the Affymetrix SNP 6.0 array.

To increase the power for eQTL discovery, we can impute variants for all samples in each cancer type using IMPUTE2 [21], with 1000 Genomes Phase 3 [24] as the reference panel. IMPUTE2 is used to phase observed genotypes and impute missing genotypes. The phasing step in IMPUTE2 can phase genotypes to best-guess haplotypes, and the following imputation step will infer the unobserved genotypes. Imputation with one-phased reference panel is the most commonly used genotype imputation scenario in IMPUTE2, which is a two-step procedure: pre-phasing and imputation of the phased data. The genotype file stores data with one line for each SNP. The SNP ID, RS ID of the

SNP, base-pair position of the SNP, the SNP allele A, and the alternative allele B should be the first five entries of each line.

The following commands show how to run this kind of analysis with IMPUTE2:

1. Pre-phasing

   Pre-phasing is a procedure that accelerates the imputation process by dividing it into two steps: (1) phase the study genotypes statistically; and (2) impute from the reference panel into the estimated study haplotypes.

   ./impute2 **-prephase_g** **-m** /path/to/fine.scale.recombination.map **-g** /path/to/file.containing.genotypes.in.study.cohort **-int** 1 5e6 **-Ne** 20000 **-o** /path/to/outputfile.prephasing.impute2

2. Imputation into pre-phased haplotypes

   ./impute2 **-use_prephased_g** **-m** /path/to/fine.scale.recombination.map **-h** /path/to/file.of.known.haplotypes **-l** /path/to/legend.file.about.SNPs.in.-h.file **-known_haps_g** /path/to/outputfile.prephasing.impute2.haps **-strand_g** /path/to/orientation.file.of.study.strand **-int** 1 5e6 **-Ne** 20000 **-o** /path/to/outputfile.phased.impute2.

   Comments

- **-g** <file>

  REQUIRED unless -known_haps_g provided

  Supply a file containing genotypes for the study cohort. The format of this file is described at http://www.stats.ox.ac.uk/~marchini/software/gwas/file_format.html. If not, supply a file of phased study haplotypes via the -known_haps_g option.
- **-m** <file>

  REQUIRED

  A fine-scale recombination map must be supplied for the region to be analyzed. This file includes information of physical position (in base pairs), genetic map position (in cM), and the recombination rate between the current position and the next position in the map (in cM/Mb). Reference datasets with appropriate recombination map files can be downloaded from https://mathgen.stats.ox.ac.uk/impute/impute_v2.html#reference.
- **-int**

  This parameter defines the boundaries of the region to be imputed on the current chromosome. For example, "-int 1 5e6" means analyzing physical positions 1–5,000,000.

Pre-phasing-based imputation is also possible on chromosome X. The only process that differs is the requirement to supply a sample file that includes gender information for data using the -sample_g option, and use the -chrX flag, and encode the male haploid genotypes according to the described file format.

After imputation, certain criteria can be used to select SNPs [13]: (1) imputation confidence score, INFO ≥ 0.4, (2) minor allele frequency (MAF) $\geq 5\%$, (3) SNP missing rate $<5\%$ for best-guessed genotypes at posterior probability ≥ 0.9, and (4) Hardy–Weinberg equilibrium P value $>1 \times 10^{-6}$ estimated by Hardy–Weinberg R package [25].

3.2 Gene Expression Data Collection and Processing

The gene expression profiles of tumor samples from TCGA program can be obtained from the GDC data portal (https://portal.gdc.cancer.gov/). Genes are retained when the average expression (RSEM calculated by the expectation-maximization algorithm [26]) is ≥ 1 in each cancer type. The expression values for each gene across all samples can be converted into a standard normal based on the rank, which can diminish the effects of outliers on the regression scores [13].

3.3 Covariate Selection from Genotype and Gene Expression Data

3.3.1 Smartpca to Generate Covariates in Genotype Data

Population stratification (allele frequency differences due to systematic ancestry differences) can cause fake associations with a disease. Stratification will be a serious issue in large-scale association studies to detect common genetic variants with weak effect. To remove the effect of population structure on genotype data, smartpca in the EIGENSOFT program [22] can be used to perform principal component (PC) analyses for each cancer type. The top PCs in the genotype data can be selected as covariates. Smartpca runs PC analysis on input genotype data and outputs PCs (eigenvectors) and eigenvalues. Five different input formats can be used. *See* /CONVERTF/README to use the convertf program for converting between formats.

The smartpca syntax is "../bin/smartpca -p parfile":

```
perl smartpca.perl -i $geno -a $snp -b $ind -k 5 -o $pca_par_
file -p $plot -e $eval -l $log
```

Various messages are printed to standard output by the smartpca program. To redirect this information to a file, adjust the above syntax to "../bin/smartpca -**p** parfile >logfile".

Parameters in parfile for smartpca are briefly described as follows:

genotypename: input genotype file
snpname: input snp file
indivname: input indiv file
evecoutname: output file of eigenvectors
evaloutname: output file of all eigenvalues

3.3.2 PEER to Generate Covariates in Expression Data

Technical confounding factors such as batch effects can complicate eQTL analysis by causing spurious associations between genetic loci and a large number of transcripts [27]. To eliminate the concealed batch effects and other confounders on gene expression, probabilistic estimation of expression residuals (PEER) software [23] can be used to generate PEER factors from expression data as covariates. PEER is a package for applying statistical models in population-scale expression data to improve the sensitivity and interpretability of genetic associations. It takes transcript profiles and covariates from a group of individuals as input, and then outputs hidden factors that interpret much of the expression variability. The inferred factors can be used in genetic association analyses. To eliminate the latent effects of clinical status on gene expression, age [28], gender [13], and tumor stage [15] should be taken as additional covariates.

The procedures for learning hidden determinants from gene expression using PEER are as follows:

1. Load PEER:

```
> library(peer)
```

2. Build the model:

```
> model=PEER()
```

3. Load the prepared data matrices:

```
> expr = as.data.frame(read.table("path of your data",
header=T, sep="\t", row.names = 1))
> header = colnames(expr)
> expr = t(expr)
```

4. Set the maximum number (e.g., 15) of unobserved factors to model.

```
> PEER_setNk(model, 15)
```

   If previous information on the magnitude of the confounding effects is unavailable, using 25% of the number of subjects enrolled in the study (but no more than 100 factors) is recommended.

5. Set the expression data.

```
> PEER_setPhenoMean(model, as.matrix(expr))
```

6. Train the model, observing convergence:

```
> PEER_update(model)
```

   If the model does not converge and the variance of the residuals keeps decreasing after the default 1000 iterations, a higher value of iterations, e.g., 10,000, can be tried (in our study, 100 iterations was sufficient [19]).

```
> PEER_setNMax_iterations(model, 10000)
```

7. Obtain the posterior mean of the inferred confounders, their weights, precision (inverse variance) of the weights, and the residual dataset:

```
> factors = PEER_getX(model)
> weights = PEER_getW(model)
> precision = PEER_getAlpha(model)
> residuals = PEER_getResiduals(model)
> write.table(cbind(header_of_exprdata, factors))
```

3.4 Identification of eQTLs

We can use the Matrix eQTL package [20] to perform eQTL analysis in a linear regression model. Matrix eQTL is software designed for calling eQTLs in large datasets with fast performance. Matrix eQTL supports models with covariates, including models with correlated and heteroskedastic errors. For each cancer type, the genotype data, expression data, and covariates should be processed to three N (genotype, expression, or covariates) $\times$ S (samples) matrix files with the sample order matched. The gene location (hg19) can be downloaded from Genomic Data Commons (https://gdc.cancer.gov/). The SNP location (hg19) can be downloaded from dbSNP (https://www.ncbi.nlm.nih.gov/projects/SNP/). eQTLs are defined as SNPs with false discovery rate (FDR) < 0.05. Cis-eQTLs and trans-eQTLs are differentiated according to the distance from the gene transcriptional start site (TSS) to the SNP, with cis-eQTLs having the SNP within 1 Mb from the TSS [13] and trans-eQTLs having the SNP beyond that point.

1. Installing and loading Matrix eQTL

```
> install.packages("MatrixEQTL")
> library("MatrixEQTL");
```

2. Set the parameters and names of the genotype and expression data files.

```
> base.dir = your_data_path
```

3. Set the parameters of the model selected and file names of the genotype and expression data.

```
> useModel = modelLINEAR; # modelANOVA or modelLINEAR or modelLINEAR_CROSS
> SNP_file_name = paste(base.dir, "/your_SNP_file_name", sep="");
> snps_location_file_name = paste(base.dir, "/your_snps_location_file_name",
sep="");
> expression_file_name = paste(base.dir, "/your_expression_file_name",
sep="");
> gene_location_file_name = paste(base.dir, "/your_gene_location_file_name",
sep="");
```

4. A separate file is provided for extra covariates.

```
> covariates_file_name = paste(base.dir, "/your_covariates_
file_name", sep="");
```

5. Set the p value threshold.

   Gene-SNP associations saved in the output file are determined by the p value threshold. The threshold should be lower for larger datasets, or this may cause excessively large output files.

```
pvOutputThreshold_cis = 1e-2;
pvOutputThreshold_tra = 1e-6;
```

6. Set the distance for local gene-SNP pairs.

```
cisDist = 1e6;
```

7. Set the path and names of the output files.

```
output_file_name_cis = "../your_output_file_name_cis";
output_file_name_tra = "../your_output_file_name_tra";
```

8. Set the parameters of the file loading.

   This section of codes includes three similar parts: loading files with genotype, gene expression, and covariates, respectively. You can set the file, the string representation for missing values, the number of rows with column labels, and the number of columns with row labels in each part. The number of variables in a slice can also be changed for the file reading procedure.

```
> snps = SlicedData$new();
> snps$fileDelimiter = "\t";      # the TAB character
> snps$fileOmitCharacters = "NA"; # denote missing values;
> snps$fileSkipRows = 1;          # one row of column labels
> snps$fileSkipColumns = 1;       # one column of row labels
> snps$fileSliceSize = 2000;      # read file in pieces of 2,000 rows
> snps$LoadFile( SNP_file_name );

> gene = SlicedData$new();
> gene$fileDelimiter = "\t";      # the TAB character
> gene$fileOmitCharacters = "NA"; # denote missing values;
> gene$fileSkipRows = 1;          # one row of column labels
> gene$fileSkipColumns = 1;       # one column of row labels
> gene$fileSliceSize = 2000;      # read file in slices of 2,000 rows
> gene$LoadFile(expression_file_name);

> cvrt = SlicedData$new();
> cvrt$fileDelimiter = "\t";      # the TAB character
> cvrt$fileOmitCharacters = "NA"; # denote missing values;
```

```
> cvrt$fileSkipRows = 1;          # one row of column labels
> cvrt$fileSkipColumns = 1;       # one column of row labels
> cvrt$LoadFile(covariates_file_name);
```

9. Load the data frame with information about SNP and gene locations.

```
snpspos = as.data.frame(read_tsv(snps_location_file_name,
col_names = TRUE ));

genepos = as.data.frame(read_tsv(gene_location_file_name,
col_names = TRUE));
```

10. Call cis- and trans-eQTL analysis.

```
me = Matrix_eQTL_main(
snps = snps,
gene = gene,
cvrt = cvrt,
output_file_name = output_file_name_tra,
pvOutputThreshold = pvOutputThreshold_tra,
useModel = useModel,
errorCovariance = errorCovariance,
verbose = TRUE,
output_file_name.cis = output_file_name_cis,
pvOutputThreshold.cis = pvOutputThreshold_cis,
snpspos = snpspos,
genepos = genepos,
cisDist = cisDist,
pvalue.hist = "qqplot",
min.pv.by.genesnp = FALSE,
noFDRsaveMemory = FALSE)
```

Comments

- cisDist—maximum distance at which a gene-SNP pair is defined as local.
- snpspos—data frame with information for SNP locations. *See* http://www.bios.unc.edu/research/genomic_software/Matrix_eQTL/Sample_Data/snpsloc.txt.
- genepos—data frame with information for gene locations. *See* http://www.bios.unc.edu/research/genomic_software/Matrix_eQTL/Sample_Data/geneloc.txt.
- genotype, expression, and covariates—their columns must correspond to samples and their rows with gene/SNP/covariate in each line. The order of the columns in all three files must match. All measurements have to be numeric and the genotype values do not have to be discrete.

- Two output files are generated for cis-eQTLs and trans-eQTLs, respectively. Each significant gene-SNP correlation is documented in a separate line, both in the output file and the returned object "*me*." Each record includes a transcript name, a SNP name, t- or F-statistic, estimation of the effect size, *p* value, and FDR.
- Matrix eQTL is designed to deal with large data sets loaded using SlicedData classes, which store the data in slices of 1000 rows (default size). The analysis will be performed for each pair of slices of genotype and expression data sets. It obtains efficiency by carrying out the most computationally intensive part of the calculations with large matrix operations, most importantly—matrix multiplications. Thus, Matrix eQTL can outperform other eQTL software by several orders of magnitude using fast BLAS in R, Revolution R, or Matlab [20]. However, fast BLAS is not included in the standard installation of R. The performance of large matrix multiplication in R can be improved up to 20 times by using a nonstandard BLAS. The necessary steps vary based on the platform. *See* http://www.bios.unc.edu/research/genomic_software/Matrix_eQTL/.

3.5 Survival-Associated eQTLs

The clinical information of tumor samples from TCGA program can be obtained from the GDC data portal (https://portal.gdc.cancer.gov/). Many genes are implicated in cancer prognosis [29], and eQTLs may affect the prognoses by altering the expression of genes. Associations between eQTLs and patients' overall survival times can be explored to identify survival-associated eQTLs. For each eQTL, samples are sorted into three groups according to genotypes: AA, Aa, and aa (A and a represent two alleles of one SNP). We can use the log-rank test to examine the differences in survival time, and Kaplan–Meier curves to represent the survival time for each group. eQTLs with FDR of log-rank test <0.05 can be defined as survival-associated eQTLs.

References

1. Cheung VG, Conlin LK, Weber TM et al (2003) Natural variation in human gene expression assessed in lymphoblastoid cells. Nat Genet 33:422–425. https://doi.org/10.1038/ng1094
2. Montgomery SB, Sammeth M, Gutierrez-Arcelus M et al (2010) Transcriptome genetics using second generation sequencing in a Caucasian population. Nature 464:773–777. https://doi.org/10.1038/nature08903
3. Pickrell JK, Marioni JC, Pai AA et al (2010) Understanding mechanisms underlying human gene expression variation with RNA sequencing. Nature 464:768–772. https://doi.org/10.1038/nature08872
4. Alemu EY, Carl JW, Bravo HC, Hannenhalli S (2014) Determinants of expression variability. Nucleic Acids Res 42:3503–3514. https://doi.org/10.1093/nar/gkt1364
5. Visscher PM, Wray NR, Zhang Q et al (2017) 10 years of GWAS discovery: biology, function, and translation. Am J Hum Genet 101(1):5–22
6. Khurana E, Fu Y, Chakravarty D et al (2016) Role of non-coding sequence variants in cancer. Nat Rev Genet 17:93–108

7. Chen Y, Zhu J, Lum PY et al (2008) Variations in DNA elucidate molecular networks that cause disease. Nature 452:429–435. https://doi.org/10.1038/nature06757
8. Cookson W, Liang L, Abecasis G et al (2009) Mapping complex disease traits with global gene expression. Nat Rev Genet 10 (3):184–194
9. Emilsson V, Thorleifsson G, Zhang B et al (2008) Genetics of gene expression and its effect on disease. Nature 452:423–428. https://doi.org/10.1038/nature06758
10. Ongen H, Brown AA, Delaneau O et al (2017) Estimating the causal tissues for complex traits and diseases. Nat Genet 49:1676–1683. https://doi.org/10.1038/ng.3981
11. Lappalainen T, Sammeth M, Friedländer MR et al (2013) Transcriptome and genome sequencing uncovers functional variation in humans. Nature 501:506–511. https://doi.org/10.1038/nature12531
12. Liang L, Morar N, Dixon AL et al (2013) A cross-platform analysis of 14,177 expression quantitative trait loci derived from lymphoblastoid cell lines. Genome Res 23:716–726. https://doi.org/10.1101/gr.142521.112
13. Ardlie KG, DeLuca DS, Segrè AV et al (2015) The genotype-tissue expression (GTEx) pilot analysis: multitissue gene regulation in humans. Science 348:648–660. https://doi.org/10.1126/science.1262110
14. Li Q, Seo JH, Stranger B et al (2013) Integrative eQTL-based analyses reveal the biology of breast cancer risk loci. Cell 152:633–641. https://doi.org/10.1016/j.cell.2012.12.034
15. Ongen H, Andersen CL, Bramsen JB et al (2014) Putative cis-regulatory drivers in colorectal cancer. Nature 512:87–90. https://doi.org/10.1038/nature13602
16. Westra HJ, Peters MJ, Esko T et al (2013) Systematic identification of trans eQTLs as putative drivers of known disease associations. Nat Genet 45:1238–1243. https://doi.org/10.1038/ng.2756
17. Zhang X, Gierman HJ, Levy D et al (2014) Synthesis of 53 tissue and cell line expression QTL datasets reveals master eQTLs. BMC Genomics 15:532. https://doi.org/10.1186/1471-2164-15-532
18. Yao C, Joehanes R, Johnson AD et al (2017) Dynamic role of trans regulation of gene expression in relation to complex traits. Am J Hum Genet 100:571–580. https://doi.org/10.1016/j.ajhg.2017.02.003
19. Gong J, Mei S, Liu C et al (2018) PancanQTL: systematic identification of cis-eQTLs and trans-eQTLs in 33 cancer types. Nucleic Acids Res 46:D971–D976. https://doi.org/10.1093/nar/gkx861
20. Shabalin AA (2012) Matrix eQTL: ultra fast eQTL analysis via large matrix operations. Bioinformatics 28:1353–1358. https://doi.org/10.1093/bioinformatics/bts163
21. Howie BN, Donnelly P, Marchini J (2009) A flexible and accurate genotype imputation method for the next generation of genome-wide association studies. PLoS Genet 5: e1000529. https://doi.org/10.1371/journal.pgen.1000529
22. Price AL, Patterson NJ, Plenge RM et al (2006) Principal components analysis corrects for stratification in genome-wide association studies. Nat Genet 38:904–909. https://doi.org/10.1038/ng1847
23. Stegle O, Parts L, Piipari M et al (2012) Using probabilistic estimation of expression residuals (PEER) to obtain increased power and interpretability of gene expression analyses. Nat Protoc 7:500–507. https://doi.org/10.1038/nprot.2011.457
24. Auton A, Abecasis GR, Altshuler DM et al (2015) A global reference for human genetic variation. Nature 526:68–74
25. Graffelman J (2015) Exploring diallelic genetic markers: the HardyWeinberg Package. J Stat Softw 64:1–23. https://doi.org/10.18637/jss.v064.i03
26. Li B, Dewey CN (2014) RSEM: accurate transcript quantification from RNA-seq data with or without a reference genome. In: Bioinformatics: the impact of accurate quantification on proteomic and genetic analysis and research. CRC Press, Boca Raton, FL, pp 41–74
27. Hyun MK, Ye C, Eskin E (2008) Accurate discovery of expression quantitative trait loci under confounding from spurious and genuine regulatory hotspots. Genetics 180:1909–1925. https://doi.org/10.1534/genetics.108.094201
28. Grundberg E, Small KS, Hedman ÅK et al (2012) Mapping cis-and trans-regulatory effects across multiple tissues in twins. Nat Genet 44:1084–1089. https://doi.org/10.1038/ng.2394
29. Gentles AJ, Newman AM, Liu CL et al (2015) The prognostic landscape of genes and infiltrating immune cells across human cancers. Nat Med 21:938–945. https://doi.org/10.1038/nm.3909

Chapter 14

QTL Analysis Beyond eQTLs

Jia Wen, Conor Nodzak, and Xinghua Shi

Abstract

Expression quantitative trait locus (eQTL) analysis is a powerful method to understand the association between genetic variant and gene expression; it also has potential impact for the study of transcription medicine for human complex disease. In the past two decades, the researchers focus on studying the eQTL, while more and more evidence shows that the regulatory genetic variants locating noncoding region have strong effect for the gene expression. More and more researchers working on eQTL analysis realize the importance of other types of QTLs beyond eQTL. In this chapter, we will explore some QTLs beyond eQTLs that show the regulatory association with eQTLs and explain the underlying link among these types of QTLs.

Key words Gene expression, eQTL, Association mapping, Molecular trait

1 Introduction

The impact of gene variants on gene expression, which is studied through eQTL analysis, is well established for human, animals, and plant. As for expression QTL, it denotes the segment of DNA that regulates the gene transcription and expression and other regulatory traits that contribute to a specific phenotypic trait [1]. In eQTL analysis, expression levels are treated as quantitative traits, and the genetic variants are treated as features. The gene expression traits are mapped to the genetic loci by the combination of the variation in gene expression patterns with genome-wide genotyping using many well-established linkage and association study tools [2]. There are two kinds of expressed quantitative trait loci, one is *cis*-eQTL which locates within specific window size away from genes, and the other is *trans*-eQTL which locates outside of specific window size away from genes. Researchers can explore the genetic architecture of gene expression variation through eQTL analysis, and then expand our understanding of regulatory functions of genetic variant on gene expression using different cell types and tissues. Finally, researchers can investigate the genetic variation on

Xinghua Mindy Shi (ed.), *eQTL Analysis: Methods and Protocols*, Methods in Molecular Biology, vol. 2082,
https://doi.org/10.1007/978-1-0716-0026-9_14,

the molecular and/or phenotype consequences through genome-wide mapping of eQTL analysis [2]. However, studies find that some proportion of eQTL have a primary effect on the chromatin, and other eQTLs enriched in the transcribed regions [3]. Thus, recently more and more studies expand to study the impact of genetic variants on post-transcriptional mechanisms, such as general RNA decay (RNA decay [rdQTLs]), alternative polyadenylation, methylation (meQTL), miRNA binding (miRNA binding QTLs [mirQTLs]), and RNA splicing QTL (sQTL) [4]. It's been verified that rdQTLs and mirQTLs are to be associated with variation in gene expression levels, sQTLs are found to be near the 5′-UTR, transcription starting sites (TSSs) and transcript binding region which can regulate the gene expression levels [5] (Fig. 1). It also finds that genetic variants strongly associated with miRNAs can increase the variation in gene expression in two studies [5, 6]. Many studies show that the post-transcriptional mechanisms can affect genetic variation in mRNA expression levels. Kilpinen et al. studied the genetic variation related with the allelic specific histone modifications using two parents-child trios and demonstrated that the key role of histone modification as one of regulatory events [7]. Kasowski et al.'s study also showed that the inherited alteration of genetic structure on the histone modification contributed to the phenotype difference through applying the RNA-seq and CHI-seq technologies on 19 individuals [8]. These studies demonstrate that it's necessary and urgent to study the regulatory variants underlying the molecular phenotype. QTL analysis is a method that can link geentic variants to various pheontypes including molecular and physical pheontypes. In eQTL analysis, the association between the genetic variants and mRNA expression levels is studied, and as for other types of QTL analysis, using protein abundance, ribosome occupancy, DNA methylation, and histone marks that are not only the transcript abundance (mRNA expression) in QTL analysis, which is linking the genetic variants to these molecular phenotypes are considered as rQTL, pQTL, meQTL, and hQTL analysis (Fig. 1).

However, the mechanisms underlying how the genetic variants affect the regulatory traits is still complex which might indicate the direct link between transcribed level traits and genetic variants. Here we introduce some types of QTL beyond eQTL including ribosome QTL (rQTL), protein abundance QTL (pQTL), histone marks (hQTL) associated with chromatin state, methylation QTL (meQTL), and others. By identifying the localized QTLs between eQTLs and other types of QTLs, Li et al. used the Bayesian hierarchical model to estimate the contribution of chromatin-level variation to the phenotype [3].

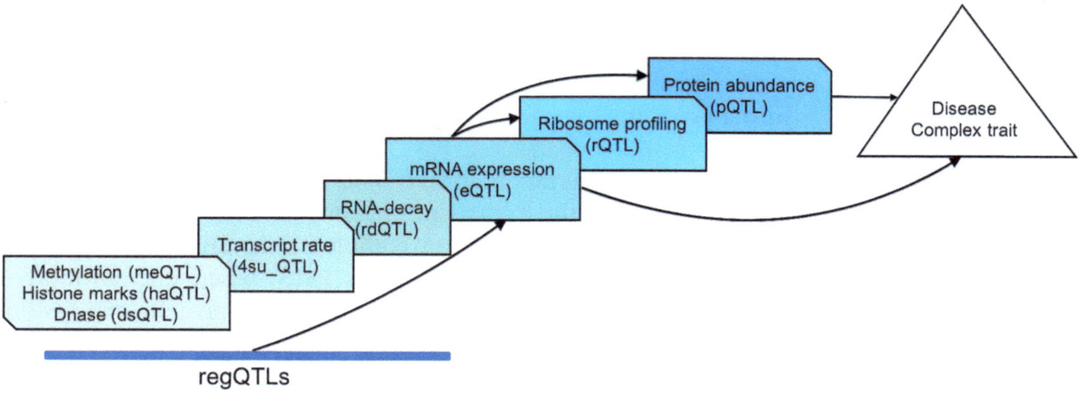

Fig. 1 The relationship between eQTL and other types of QTLs and phenotype

2 QTLs Beyond Other Types of eQTLs

2.1 RegQTLs Introduction

Regardless of the transcription factor binding mechanism, it is obvious that not all the regulation actives in the promoter region, the underlying fact is that promoters and enhancers perform beyond the known post-transcription effect. Hence, researchers study the regQTLs to find interaction between regulatory elements. The regQTL means that the genetic variants show the association across the mRNA expression and other molecular traits, such as methylation level, histone marks level and other chromatin traits. For example, the sQTLs and rdQTLs which are mainly enriched in the splice sites, 3′ UTR motif region and transcript starting sites, and they additionally have regulatory effect on the gene expression together with regulatory mechanism. Hence, it highlights a fact that it's not separate for these different types of QTLs. Complex co-regulatory mechanism is relatively common in the biology identity. Moreover, few regQTLs show much effect on the gene expression although studies on regQTLs improve our understanding for the biological mechanism of eQTLs. For instance, one study showed that 55% eQTLs are identified dsQTLs, while only 39% dsQTLs were found to be associated with gene expression [5]. Combined analysis of quantitative post-transcription, ribosome profiling and protein measurements provides understandable insight for the regQTLs which are also eQTLs, and non-eQTLs regQTLs [5].

2.2 Definition of pQTL

It is considered that genome variants associated with mRNA expression variation can be associated with protein-level variation that impacts complex traits [9]. Therefore, protein quantitative trait loci (pQTLs) are defined to be genetic variants that are associated with protein abundance levels (pQTLs) [10]. Since most eQTLs can affect the mRNA levels, the finaly consquential effects on traits or pheontype can be mediated through ribosome profiling levels and protein abundance levels [10, 11]. Genetic variation can affect the expression of mRNA transcripts; thus, it will have a profound effect on the amount of mRNA available to translate into a protein.

2.3 Current Research on pQTL's Analysis

Consequently, genetic variation can have effect on the expression of protein through the transcriptional mechanism and will consequently affect the downstream complex traits and phenotype. In protein analysis, mass spectrometers (MS)-based and antibody-based method can be used in assessing the protein expression level [9, 12]. However, mRNA expression variation cannot perfectly show the variation of protein [4]. Because the protein abundance would be affected by post-transcriptional modification, many studies have demonstrated that the relationship between the mRNA expression and protein abundance is frequently modest, and furthermore the protein abundance is inheritable and can be affected by genetic variants [2, 9, 13–15]. Wu et al. used the HapMap populations to study the protein abundance variation and found that the protein abundance levels were inheritable and could be considered as one molecular trait to be studied using QTL methods [15]. Foss et al. studied the genetic variants in both protein abundance and mRNA expression and compared the pQTL result with eQTL result in a cross by two diverse strains of yeast. The result from this research showed that the pQTLs were not the same with eQTLs and demonstrated the importance of directly studying the genetic variants associated with proteome [16]. Battle et al. made the first integrated genetic research combining the cascade hierarchical of mRNA levels, ribosome profiling, and protein abundance in human QTL analysis [10]. Hence, it's necessary to study the relationship between the genome and proteome. Protein QTL (pQTL) denotes the genetic variants with effect on the protein abundance in proteome-wide analysis. The pQTL analysis links the genetic variants to the protein expression, and intents to identify the genetic variants that impact the protein level. The genetic variants can be SNPs and/or structure variation (SV). The figure below illustrates the molecular quantity of different biological progress for each sample [17] (Fig. 1). The gene expression cascade structure can be studied in the same way as eQTL (using mRNA expressions); we use the DNA methylation (me), histone modifications, transcription factor (TF) binding, active transcription, mRNA levels (resulting in eQTLs), translation, and protein levels (resulting in protein QTLs (pQTLs) to study the associated genetic variants. In the Fig. 1, the genetic variants can affect the mRNA levels, then the protein abundance, and finally affect high or low risk of the disease. Many statistical methodologies can be applied in analyzing pQTL like eQTL analysis. We summarize several commonly used methods in the Table 1 [18]; these QTL mapping methods are given detailed description in

Table 1
Summary of methods used in QTL analysis [18]

Tools	Reference	Website available
WGCNA	[19]	https://www.bioconductor.org/
MatrixEqtl	[20]	http://www.bios.unc.edu/research/genomic_software/Matrix_eQTL/
WASP	[21]	https://github.com/bmvdgeijn/WASP
SPIRE	[22]	https://bitbucket.org/bereste/spire
FM-eQTL	[23]	https://github.com/xqwen/fmeqtl
eQTLBMA	[23–25]	https://github.com/timflutre/eqtlbma

another chapter titled as "Introductory Methods for eQTL Analyses."

Those methods are developed for identifying genetic-variants QTLs. Many studies have made efforts to identify genetic variants associated with protein expression in yeast, mouse brain tissue, and human [10, 16, 18, 26, 27]. Melzer et al. used European population to find eight *cis*-effect gene variants associated with proteins abundance levels involved in human diseases [28]. However, the proteome QTL analysis lags the eQTL analysis and few reports focus on the protein-QTL. In recent years, the most challenging point is the methodology development due to the high-scale quantification of proteomics. Several methods in Table 1 could be used to solve the protein-QTL model like eQTLs.

2.4 Definition of rQTL

Ribosome profiling is another molecular trait in biological progress and can reflect the translation efficiency from mRNA abundance through ribosome profiling [14, 29]. Ribosome regulates the translation progress which plays a pivot role from gene expression to the phenotype. Ribosome profiling can largely increase the precision of assessing the protein abundance prediction. Therefore, we have another type of QTL, ribosome occupancy (rQTL) beyond eQTL and pQTL. Ribosome occupancy QTL can be considered as genetic variants associated with ribosome occupancy measured by ribosome profiling.

2.5 Current Research on rQTL's Analysis

Several studies have reported rQTLs studied by the method of using the ribosome profiling as the expression levels in yeast and human, and results from these studies show that the rQTL are more concordant with eQTL results than pQTLs, which means that the protein might be affected by post-translational modification. Alert et al. examined the ribosome profiling of a cross by two strains of yeast and found that the transcripts differences were highly correlated with ribosome abundance [17]. In Battle et al. study, 72 YRI LCLs were used for integrating genetic analysis; the result showed that around 90% of ribosome occupancy QTLs (rQTLs) were associated with mRNA expression levels, while only 35% of eQTLs affect the variation of protein abundance levels [10]. The result shows that the correlation between eQTLs and rQTLs is higher than with pQTLs, which means that the protein variation may be due to the post-translational or post-transcription modification. The overlap between the rQTLs and pQTLs also demonstrates that the genetic variants can affect the protein expression directly through affecting the translational levels.

2.6 Definition of hQTL

Similarly, linking genetic variants to the histone modification marks or other epigenetic marks of regulatory elements such as promoters or enhancers variation is histone QTL (hQTL) analysis.

2.7 Current Research on hQTL Analysis

Many human complex diseases are caused by the genetic variants located in non-coding region across chromosomes, and illustrating their functions is an urgent task for researchers. Histone modification marks measured by Chip-Seq can reveal the chromatin state which are associcated with accessible chromatin regions; thus, they can alter the gene expression by the effect of the chromatin structure at the post-translational stage. Hence, studying the hQTLs could help to understand how the genetic regulatory variation affects gene expression in human. Several studies report that hQTL are enriched with variants associated immune disease in human. McVicker et al. used four histone marks that are promoters and enhancers to identify hQTL in ten YRI LCLs and found that most of them overlap with dsQTL [30]. Grubert et al. identified hQTLs using three histone marks that are H3K27ac, H3K4me, and H3K4me3 in LCLs across 75 samples, and results demonstrate that hQTLs are enriched with SNPs associated with autoimmune human disease. They also found that 66% eQTL overlapped with hQTLs [13].

2.8 Definition of rdQTL

rdQTL denotes the genetic variants associated with RNA decay rate. In eQTL analysis, gene expression which measures the steady-state of transcription levels can't be used for distinguishing the variation existing in steady-state gene expression levels and RNA decay rates. Moreover eQTL analysis is well-established, while RNA decay rate QTL identification lags behind eQTL analysis [5].

2.9 Current Research on rdQTL Analysis

According to Pai et al., most rdQTLs showed association with higher steady-state expression, and also showed association with higher rapid RNA decay rates [5]. The mechanism of the opposite effect for common sense between RNA decay rate and mRNA expression is still unclear, which means that the rdQTLs are complex in other side. However, it indicates the buffering mechanism between the transcription and RNA decay rate. Pai et al. showed that nearly 20% of eQTLs are driven by the different RNA decay rate [5].

2.10 Definition of meQTL

MeQTLs are methylation quantitative trait loci, which is similar to other types of QTLs that affect the DNA methylation level (typically CpG methylation). In meQTLs analysis, DNA methylation level is viewed as the quantitative trait, and the genetic variants are viewed as the features in order to identify the meQTLs. The variation of DNA methylation levels occurring at regulating regions can affect the gene expression [31].

2.11 Current Research on meQTL Analysis

In epigenetics, DNA methylation that is related to regulatory of gene expression is proven to be an important epigentic mark. Studies provide evidence that the quantitative variation of DNA methylation is associated with genetic variation. meQTLs are known to be associated with multiple molecular pheontypes such as gene expression, DNase I hypersensitivity, PolII occupancy, and several histone marks suggesting that they are envolved in multiple molecular regulatory events [32]. The quantitative variation of transcription factor and the transcription factor abundance are highly correlated with the DNA methylation variation which locates near the transcript starting sites (TSSs) [32]. Banovich et al. conducted meQTL analysis using 64 YRL LCLs from HapMap project and found that meQTLs usually were associated with the variation of transcription factor binding, histone modification, and mRNA expression levels [32]. The genome-wide variation on the diseases or cancer can be obtained through GWAS [31, 33], meQTLs study can help to further extend GWAS information, and additionally the DNA methylation phenomenon is usually observed in the cancer epigenetic study [31].

2.12 Definition of dsQTL

DNase I sequence technology can be used to measure the genome-wide chromatin accessibility and can provide the genome-wide map for chromatin accessibility in individuals. Upregulated DNase I is a symbol of open chromatin, and DNase I hypersensitive sites (DHSs) can be used to mark the region where the active histone marks and transcript factors combine [30]. The variation of chromatin accessibility and transcript factors binding can occur at some gene loci which are considered to largely affect the phenotypic variation [1]. Inserting/deleting the DNase I sensitivity quantitative trait loci (dsQTL) nearby genes can cause the variation of mRNA expression levels, which means that the DNase correlates with the nearby genes.

2.13 Current Research on dsQTL Analysis

Degner et al. conducted joint eQTL and dsQTL analysis using genome-wide gene expression and genome-wide genotypes for 70 Yoruba lymphoblastoid cell lines (LCLs). In this study, authors overlapped the dsQTLs and eQTLs and found ~16% of dsQTL can affect the changes of nearby gene expression, while ~55% of eQTLs can affect the nearby DNase sensitive levels. Besides that, results from this study showed that most dsQTLs were enriched within TSSs regions and could induce the allele specific phenomenon in transcript factor binding regions [1]. The joint dsQTL-eQTL analysis revealed that dsQTLs are a major dominant factor in gene regulation and gene expression variation [34].

2.14 Definition of sQTL

sQTL means the genetic variants that affect gene regulation through the way of pre-mRNA splicing events, which can be measured by gene expression and the amino acid sequences of the resulting proteins [3]. Actually, more than 90% of genes are alternatively spliced [3, 35].

2.15 Current Research on sQTL

The splicing events can induce the diversity of mRNA, which process can produce different proteins from the same genomic sequences [36]. The sQTLs would affect the gene regulation and function through pre-RNA splicing events [3]. A good understanding on the splicing events can help prevent the aberrant proteins and will shed light on the development of novel targets for cancers [37]. Hence, the sQTL analysis usually adopts the RNA-seq which can statistically measure the isoform-specific gene expression levels. The isoform-specific gene expression levels can be viewed as quantitative traits, and similarly used the genotype variants as the features in sQTL analysis. Li et al. systematically investigated the effect of genetic variants on chromatin levels, RNA splicing events, ribosome and protein levels and built up a novel method to measure the splicing variation in order to identify splicing QTLs [3]. The splicing QTLs and eQTLs play comparable role in regulating gene expression and post-transcriptional mechanism according to this study. According to the study, sQTLs were enrichedf within gene bodies including exons and introns while eQTLs enriched in TSSs and GWAS risk loci. Hence, it's considered that eQTLs and sQTLs might be independent with each other [3].

3 Conclusion

In this chapter, we reviewed QTLs beyond eQTLs, including pQTLs, rQTLs, hQTLs, rdQTLs, meQTLs, and sQTLs, and illustrated the relationship between QTLs and eQTLs. Genetic variation can not only affect the gene expression variation, chromatin-level variation, and but also affect gene regulation and function through pre-RNA splicing. The study of combination of eQTLs and other types of QTLs would help to link the get better understanding on how genetic variants affect complex disease and quantitative traits. Furthermore, combining types of QTLs studies and the genome-wide association studies can help to identify the genetic variants as potentially GWAS causal hits. Variation in all molecular traits in the QTL analysis could impact the gene expression levels, the gene regulation changes, and finally the phenotypes of organism.

References

1. Degner JF, Pai AA, Pique-Regi R, Veyrieras JB, Gaffney DJ, Pickrell JK et al (2012) DNase I sensitivity QTLs are a major determinant of human expression variation. Nature 482 (7385):390
2. Gilad Y, Rifkin SA, Pritchard JK (2008) Revealing the architecture of gene regulation: the promise of eQTL studies. Trends Genet 24 (8):408–415
3. Li YI, van de Geijn B, Raj A, Knowles DA, Petti AA, Golan D et al (2016) RNA splicing is a primary link between genetic variation and disease. Science 352(6285):600–604
4. Schwanhäusser B, Busse D, Li N, Dittmar G, Schuchhardt J, Wolf J et al (2011) Global quantification of mammalian gene expression control. Nature 473(7347):337
5. Pai AA, Pritchard JK, Gilad Y (2015) The genetic and mechanistic basis for variation in gene regulation. PLoS Genet 11(1):e1004857
6. Pai AA, Cain CE, Mizrahi-Man O, De Leon S, Lewellen N, Veyrieras JB et al (2012) The contribution of RNA decay quantitative trait loci to inter-individual variation in steady-state gene expression levels. PLoS Genet 8(10): e1003000
7. Kilpinen H, Waszak SM, Gschwind AR, Raghav SK, Witwicki RM, Orioli A et al (2013) Coordinated effects of sequence variation on DNA binding, chromatin structure, and transcription. Science 342 (6159):744–747
8. Kasowski M, Kyriazopoulou-Panagiotopoulou S, Grubert F, Zaugg JB, Kundaje A, Liu Y et al (2013) Extensive variation in chromatin states across humans. Science 342(6159):750–752
9. Hause RJ, Stark AL, Antao NN, Gorsic LK, Chung SH, Brown CD et al (2014) Identification and validation of genetic variants that influence transcription factor and cell signaling protein levels. Am J Hum Genet 95 (2):194–208
10. Battle A, Khan Z, Wang SH, Mitrano A, Ford MJ, Pritchard JK, Gilad Y (2015) Impact of regulatory variation from RNA to protein. Science 347(6222):664–667
11. de Sousa Abreu R, Penalva LO, Marcotte EM, Vogel C (2009) Global signatures of protein and mRNA expression levels. Mol BioSyst 5 (12):1512–1526
12. Johansson Å, Enroth S, Palmblad M, Deelder AM, Bergquist J, Gyllensten U (2013) Identification of genetic variants influencing the human plasma proteome. Proc Natl Acad Sci U S A 110(12):4673–4678
13. Grubert F, Zaugg JB, Kasowski M, Ursu O, Spacek DV, Martin AR et al (2015) Genetic control of chromatin states in humans involves local and distal chromosomal interactions. Cell 162(5):1051–1065
14. Ingolia NT, Ghaemmaghami S, Newman JR, Weissman JS (2009) Genome-wide analysis in vivo of translation with nucleotide resolution using ribosome profiling. Science 324 (5924):218–223
15. Wu L, Candille SI, Choi Y, Xie D, Jiang L, Li-Pook-Than J et al (2013) Variation and genetic control of protein abundance in humans. Nature 499(7456):79
16. Foss EJ, Radulovic D, Shaffer SA, Ruderfer DM, Bedalov A, Goodlett DR, Kruglyak L (2007) Genetic basis of proteome variation in yeast. Nat Genet 39(11):1369
17. Albert FW, Kruglyak L (2015) The role of regulatory variation in complex traits and disease. Nat Rev Genet 16(4):197
18. Nédélec Y, Sanz J, Baharian G, Szpiech ZA, Pacis A, Dumaine A et al (2016) Genetic ancestry and natural selection drive population differences in immune responses to pathogens. Cell 167(3):657–669
19. Langfelder P, Horvath S (2008) WGCNA: an R package for weighted correlation network analysis. BMC Bioinformatics 9(1):559
20. Shabalin AA (2012) Matrix eQTL: ultra fast eQTL analysis via large matrix operations. Bioinformatics 28(10):1353–1358
21. Van De Geijn B, McVicker G, Gilad Y, Pritchard JK (2015) WASP: allele-specific software for robust molecular quantitative trait locus discovery. Nat Methods 12(11):1061
22. Kel I, Chang Z, Galluccio N, Romeo M, Beretta S, Diomede L et al (2016) SPIRE, a modular pipeline for eQTL analysis of RNA-Seq data, reveals a regulatory hotspot

controlling miRNA expression in *C. elegans*. Mol BioSyst 12(11):3447–3458
23. Wen X (2014) Bayesian model selection in complex linear systems, as illustrated in genetic association studies. Biometrics 70(1):73–83
24. Flutre T, Wen X, Pritchard J, Stephens M (2013) A statistical framework for joint eQTL analysis in multiple tissues. PLoS Genet 9(5): e1003486
25. Wen X, Stephens M (2014) Bayesian methods for genetic association analysis with heterogeneous subgroups: from meta-analyses to gene-environment interactions. Ann Appl Stat 8 (1):176
26. Klose J, Nock C, Herrmann M, Stühler K, Marcus K, Blüggel M et al (2002) Genetic analysis of the mouse brain proteome. Nat Genet 30(4):385
27. Garge N, Pan H, Rowland MD, Cargile BJ, Zhang X, Cooley PC et al (2010) Identification of quantitative trait loci underlying proteome variation in human lymphoblastoid cells. Mol Cell Proteomics 9(7):1383–1399
28. Melzer D, Perry JR, Hernandez D, Corsi AM, Stevens K, Rafferty I et al (2008) A genome-wide association study identifies protein quantitative trait loci (pQTLs). PLoS Genet 4(5): e1000072
29. McManus CJ, May GE, Spealman P, Shteyman A (2014) Ribosome profiling reveals post-transcriptional buffering of divergent gene expression in yeast. Genome Res 24 (3):422–430
30. McVicker G, van de Geijn B, Degner JF, Cain CE, Banovich NE, Raj A et al (2013) Identification of genetic variants that affect histone modifications in human cells. Science 342 (6159):747–749
31. Tycko B (2010) Mapping allele-specific DNA methylation: a new tool for maximizing information from GWAS. Am J Hum Genet 86 (2):109–112
32. Banovich NE, Lan X, McVicker G, Van de Geijn B, Degner JF, Blischak JD et al (2014) Methylation QTLs are associated with coordinated changes in transcription factor binding, histone modifications, and gene expression levels. PLoS Genet 10(9):e1004663
33. Gibson G, Powell JE, Marigorta UM (2015) Expression quantitative trait locus analysis for translational medicine. Genome Med 7(1):60
34. Li MJ, Yan B, Sham PC, Wang J (2014) Exploring the function of genetic variants in the non-coding genomic regions: approaches for identifying human regulatory variants affecting gene expression. Brief Bioinform 16 (3):393–412
35. Nilsen TW, Graveley BR (2010) Expansion of the eukaryotic proteome by alternative splicing. Nature 463(7280):457
36. Wang ET, Sandberg R, Luo S, Khrebtukova I, Zhang L, Mayr C et al (2008) Alternative isoform regulation in human tissue transcriptomes. Nature 456(7221):470
37. Jia C, Hu Y, Liu Y, Li M (2015) Mapping splicing quantitative trait loci in RNA-Seq: supplementary issue: sequencing platform modeling and analysis. Cancer Inform 14(Suppl 1):45–53

Chapter 15

Quantitative Trait Loci (QTL) Mapping

Kara E. Powder

Abstract

Quantitative trait loci (QTL) are genetic regions that influence phenotypic variation of a complex trait, often through genetic interactions with each other and the environment. These are commonly identified through a statistical genetic analysis known as QTL mapping. Here, I present a step-by-step, practical approach to QTL mapping along with a sample data file. I focus on methods commonly used and discoveries that have been made in fishes, and utilize a multiple QTL mapping (MQM) approach in the free software package R/qtl.

Key words Quantitative trait locus (QTL), Genetic basis, Quantitative genetics, Epistasis, MQM

1 Introduction

Quantitative trait loci (QTL) mapping is a powerful technique to genetically detect genomic intervals that contribute to complex, often continuous, phenotypes. Notably, this technique can determine gene(s) that underlie variation in an unbiased way, identify genes with varying impacts on the trait, and powerfully detect genes that have small, cumulative effects on phenotypic variation. Through identification of intervals containing a QTL (commonly termed a QTL peak), we can begin to understand (1) the number and effect size of genes (e.g., a small number of genes with major effects or many genes with minor effects each), (2) the interaction between genes (gene x gene interactions or epistasis) and genes and the environment, and (3) candidate gene(s) and nucleotides that mediate phenotypic variation. All of these factors combine to generate the so-called genetic architecture or genotype–phenotype map [1], the structure of which has major implications on the development and evolution of the trait [2].

The technique of QTL mapping builds on the principles of quantitative genetics, a field that began over a century ago [3–7]. As such, it is unfeasible to thoroughly discuss all theory and statistics that underlie QTL mapping methods. Here, I provide an

Xinghua Mindy Shi (ed.), *eQTL Analysis: Methods and Protocols*, Methods in Molecular Biology, vol. 2082,
https://doi.org/10.1007/978-1-0716-0026-9_15,

overview of the process and a step-by-step process of mapping a QTL. I also refer the reader to a number of excellent resources that discuss the history [8], statistical framework [8–10], and practical aspects of conducting QTL mapping [9, 11, 12]. These resources contain example problems and calculations [8–10], are in both book [8–10] and review [13, 14] formats, and include those that intended as introductions to the field [10].

QTL mapping can be conducted on nearly all species, lines, or strains that may be reared in the lab, exhibit sufficient phenotypic variation, and produce viable hybrids when crossed. While the rest of this work will focus on findings and methods that are most common in QTL analyses in fishes, the approach used here can be applied to many other organisms. In fact, most of the methods used to map QTL were initially established within model organisms such as *Drosophila* [15, 16], mice [17–19], and crops [20–24]. These early approaches were then extended to closely related species (e.g., sister species in *Drosophila* [25–27] or maize versus its wild ancestor teosinte [28]), and subsequently to a menagerie of nontraditional model organisms.

Within fishes, QTL mapping has been especially fruitful to understand the genetic basis of evolved differences between marine and freshwater sticklebacks, cave- and surface-dwelling *Astyanax*, and the adaptive radiation in cichlid fishes. Further, this approach has been useful in the analysis of ecologically and commercially relevant traits in aquaculture, for example, body weight and growth rate [29–34] in carp, salmon, sea bass, tilapia, and trout. QTL analysis in fishes has primarily been focused on morphological traits, likely due to ease of phenotyping. QTL have been identified for length and width of various facial bones [35–37], dentition [37–39], number of gill rakers [35, 40], size and morphology of scales [41], fin size [35, 42], thickness of the retina [43], and the number and size of armor plates [40, 44–46] and spine lengths [35, 40, 44, 47] in sticklebacks. In addition to these studies that counted or measured the length of a particular structure, QTL have also been identified for morphologies that are more complex to quantify using techniques such as statistical shape analyses [48, 49]. QTL have been identified for brain volume [50], body asymmetry [47, 51], and the overall shapes of the body [44, 52, 53], scales [41], and head structures [37, 53, 54].

QTL mapping can be conducted for any quantifiable trait. Beyond morphology, QTL mapping has been applied to understand color pattern [55, 56] and color levels [56–58], both of which are crucial for mating and sexual selection in fishes. Behavioral traits including startle response [59], schooling tendency [60], and anti-predator behaviors such as shoaling [61, 62] have all been assessed using QTL mapping in fishes. Selective breeding in aquaculture has benefitted from knowledge of the genetic basis of physiological traits such as drug or toxin tolerance [63, 64], pathogen response and resistance [65–67], stress response to crowding [68], fat deposition [62], salinity tolerance [69], time of sexual maturation [70], and gonad size [71].

QTL mapping has also been applied quantitative traits such as changes in expression of an individual or set of genes (to identify an expression QTL or eQTL [72]). In zebrafish, over 300 associations between copy number variations and gene expression have been identified [73]. QTL mapping has also identified the genetic basis of changes in gene expression following size-based harvesting regimes in fisheries [74] and adaptation to marine (saltwater) and stream (freshwater) environments [75, 76]. Finally, QTL mapping can identify the genetic loci that influence the quantitative correlations between traits (relationship QTL, or rQTL [77]). For example, genetic mapping in cichlids has been used for two measures of the covariation of traits, integration [56, 78] and modularity [79].

All these QTL approaches start by asking if there is any statistically significant difference in phenotype between individuals within each genotype group (i.e., AA, AB, and BB), and repeating this across the genome. QTL analysis is based on linkage disequilibrium between a causative genotype and genetic variation at nearby variable loci termed markers. The multiple-QTL mapping (MQM) approach used here (based on [80–82] and *see* [11]) accounts for covariation of genetic factors, reduces residual phenotypic variation, and increases the power to detect QTL; this approach is particularly appropriate for complex traits that are expected to be affected by multiple loci. MQM involves building a statistical model using cofactors. These cofactors are verified based on multiple regressions and backward elimination before inclusion in the final model. Statistical significance of a QTL is measured using a logarithm of odds (LOD) score that represents the difference of likelihoods between the alternative hypothesis (presence of a QTL) and null hypothesis (absence of a QTL).

All commands used within computer software are presented in italics, with each new line beginning with the character ">" for clarity. An electronic script of the methods and notes section compatible with R software is available at https://github.com/kpowder/MiMB_QTL.

2 Materials

2.1 Cross Generation

The choice of parental species or strains is critical. Parents must differ in terms of both phenotype of interest and genotypes (e.g., single nucleotide polymorphisms [SNPs]). Once parents are chosen, breed the two parental species or strains together to generate a heterozygous F_1 hybrid generation. F_1 hybrids will then be used to generate animals in which QTL mapping will be conducted. However, additional information can be gleaned from these animals. For example, heritability is often estimated by regression of phenotypic measures in F_1 offspring versus parents [10, 83]. Further, the number of loci underlying a trait can be estimated based on parental and hybrid phenotypic distributions using the Castle-Wright estimator ([3] and *see* modifications of the original formula in [8], pages 233–249), among others [84].

Here we will assume an F_2 cross design, in which F_1 hybrids are inbred to generate an F_2 population, as this design is the most common strategy for QTL in fishes and for the reasons below. A backcross design, in which F_1 hybrids are crossed to either parental, is also regularly used, but is less preferable for the following reasons. Backcrossing to a parent can only produce two different genotypes at each locus (i.e., AA and AB). As an F_2 cross can generate three genotypes (i.e., AA, AB, and BB) at each locus, the F_2 cross design allows assessment of more modes of action (e.g., dominant, recessive, and additive effects). Further, F_2 crosses have increased power and may only require half as many animals for QTL analysis [8].

QTL regularly need large F_2 sample sizes (i.e., hundreds of animals), but this varies based on allele frequencies, phenotypic variation within and between parents, heritability of the trait, effect size of each QTL, and genetic architecture including number and interaction of loci [10, 14]. Power and minimal sample size estimates can be conducted using the ***powercalc*** and ***samplesize*** functions, respectively, in the program R/qtlDesign [85].

2.2 Genotype Acquisition and Generating a Linkage Map

QTL mapping originally used restriction fragment length polymorphisms (RFLPs), randomly amplified polymorphic DNAs (RAPDs), and microsatellites as molecular markers. While these are still effective, genotyping at markers across the genome is increasingly utilizing Next-Generation Sequencing (NGS) approaches [86, 87]. Particularly useful in non-model organisms have been restriction-site associated DNA sequencing (RAD-seq) [88, 89] and genotype-by-sequencing (GBS) [90, 91]. These both use restriction enzymes to reduce genome complexity and can be combined with other methods such as pooling individuals [92] (termed bulk segregant analysis) to most cost-efficiently genotype individuals.

Following high-throughput DNA sequencing, short DNA reads are bioinformatically processed (e.g., trimming barcodes), reads are mapped to the genome, and variants such as single nucleotide polymorphisms (SNPs) are called. Additional filtering to produce the final set of SNPs for mapping may include excluding markers that did not meet minimum read depth thresholds (e.g., supported by >20 reads), were not differentially fixed between parents, had excessive missing genotypes (e.g., missing in >25% of hybrids), and did not demonstrate Hardy-Weinberg equilibrium and/or Mendelian segregation.

If not already available, a linkage map must be generated from marker genotypes. Using recombination frequencies, all markers will be classified into linkage groups, markers will be ordered within the linkage group, and genetic map distances will be calculated. The most commonly used software for this is JoinMap [93] or the *formLinkageGroups* and *orderMarkers* functions in R/qtl [9]. A linkage map can also be estimated as the data is loaded (*see* Subheading 3.1, **step 4**).

2.3 Phenotype Acquisition

Any quantifiable phenotype can be assessed by QTL mapping, as illustrated by the array of phenotypes described in Subheading 1. If necessary to remove the effects of allometry, measured phenotypes should be converted to residuals by normalizing to standard length or other appropriate size measurement. For morphological traits, this size correction is routine. However, for traits such as pigmentation or behavior which may not have a relationship with allometry, the effect of size can be assessed with ANOVA to see if correction is necessary. It is assumed that phenotypes have a normal distribution. If necessary, phenotypic data should be transformed to meet this assumption (e.g., log transformation).

2.4 Software

Many QTL mapping software packages are available. The method described here will utilize a commonly used and free program called R/qtl [9, 11]. This program runs within the free R software environment, which can be downloaded at http://www.r-project.org/. Many users choose to use the interface R studio; this is an extension of R (i.e., you must still download R software for R studio to work), and is available for free at https://www.rstudio.com. Once R is installed, download the R/qtl package.

```
>install.packages("qtl")
```

3 Methods

3.1 Preparing and Loading Data

1. Prepare the data as a .csv file with the correct headings (*see* **Notes 1** and **2**). Genotypes are standardly in the format "AA," "AB," and "BB." All missing phenotypes and genotypes should be noted by "NA" to work with commands below (*see* **Note 3**).
2. Once R (or R studio) is open, load the r/QTL software.

```
>library(qtl)
```

3. From the toolbar menu, select Session, Set Working Directory, then Choose Directory to navigate to the folder that contains your .csv data file.

4. Read in the genotypes, genetic map (*see* **Note 4**), and phenotypes for your mapping cross from your .csv file. In the below example, you will be prompted to select your file.

```
>data<-read.cross(format="csv", file=file.choose(),
genotypes=c("AA", "AB", "BB"), na.strings=c("NA"),
convertXdata=TRUE)
```

3.2 Data Verification

1. High-quality data is critical and errors in genotype or phenotype data can produce strange mapping results. As a first step, look at details of the data set and verify the type of cross, number of individuals, number of phenotypes (*see* **Note 5**), and genotypic data (*see* **Note 6**).

```
>summary(data)
```

2. Visually inspect your data including missing genotypes (*see* Subheading 3.2, **step 4**), the genetic map, and histogram distributions for each phenotype column.

```
>plot(data)
```

3. Visualize your mapping data (*see* **Note 7**).

```
>geno.image(data)
```

4. Identify markers and hybrids that have too many missing genotypes (*see* **Note 8**).

```
>plotMissing(data)
```

5. Remove individuals that are missing too many genotypes. The below will only retain in the data set individuals that have greater than 50 genotypes in the data set.

```
>data<-subset(data, ind=(ntyped(data)>50))
```

6. Remove markers that have too much missing data. The *ntyped* command will list the number of individuals genotyped for each marker. Any markers below a certain threshold can be removed by the *drop.markers* command and listing the name of the specific marker to drop, here marker1.

```
>ntyped(data, "mar")
>data<-drop.markers(data, "marker1")
```

7. Fill in your missing data (i.e., "NA" genotypes) (*see* **Note 9**).

```
>augdata<-mqmaugment(data,minprob=0.1)
```

3.3 Building a Statistical Model

1. Conduct an initial scan, here for the phenotypic data in column 2 (*see* **Note 10**) to identify putative unlinked QTL that will be used to build a more rigorous model (*see* **Note 11**).

```
>scan<-mqmscan(augdata,pheno.col=2,plot=T,model=
"dominance",verbose=FALSE)
```

2. View a summary of marker locations with the top LOD scores per chromosome (*see* **Note 12**) to identify putative QTL that you will use as cofactors to generate a more robust model (*see* **Note 13**).

```
>summary(scan)
```

3. For each putative QTL (*see* **Notes 14** and **15**), visually verify the effect of genotype on phenotype by generating an effect plot. The below command is for the locus on chromosome 1 at 5 centiMorgans (cM) (*chr=1, pos=5*). If at least two genotype groups do not have overlapping ranges of phenotypes, this is a putative QTL that should be added as a cofactor in your model. More details on effect plots are discussed in Subheading 3.5, **step 3**.

```
>effect<-effectplot(augdata,pheno.col=2,mname1=find.
marker(augdata, chr=1, pos=5))
```

4. Establish the file of cofactors you will use.

```
>cofactorslist<-NULL
```

5. Build your list of cofactors (*see* **Note 16**), repeating the below with new chromosome and position values for all putative QTL (*see* **Notes 17** and **18**).

```
>cofactorslist<-c(cofactorslist,find.markerindex(augdata,
find.marker(augdata,chr=1,pos=5)))
```

6. Generate the cofactor matrix (*see* **Notes 19–21**).

```
>cofactors<-mqmsetcofactors(augdata,cofactors=c
(cofactorslist))
```

7. Run a QTL scan using your updated model with cofactors. You will use the output of this to further refine your model (*see* **Note 22**).

```
>scan<- mqmscan(cross=augdata,pheno.col=2,cofactors=
cofactors,cofactor.significance=0.002, verbose=T,plot=T,
model="dominance")
 >summary (scan)
```

8. If there are any new putative QTL, assess the effect plot at this locus (*see* Subheading 3.3, **step 3**), adding these as cofactors in the model as necessary (*see* Subheadings 3.3, **steps 5** and **6**).
9. Repeat the process until the model stabilizes (*see* **Notes 23–25**): build a new model with additional cofactors, run the scan with the updated model, assess new putative QTL using effect plots, and build a new model with these updated cofactors.

3.4 Statistical Significance

1. Determine the threshold of LOD that meet the 5% and 10% statistical significance levels (i.e., $p < 0.05$ and $p < 0.10$, respectively, *see* **Note 26**) by running a permutation on your final model, here 1000 permutations (*see* **Notes 27** and **28**).

```
>result<-mqmpermutation(cross=augdata,scanfunction=
mqmscan,cofactors=cofactors,pheno.col=2,n.perm=1000,
plot=F,verbose=T,model="dominance")
 >resultqtl<-mqmprocesspermutation(result)
 >summary(resultqtl)
```

2. For each significant QTL, run Bayesian analysis for each peak to get 95% confidence interval and closest flanking markers (*see* **Note 29**). Replace the "X" in the below with your chromosome number.

```
>bayesint(scan, X, qtl.index=1,prob=0.95,lodcolumn=1, expandtomar-
kers=T)
```

3.5 Assessing QTL

1. View the LOD plot across the genome.

```
>plot(scan)
```

2. View the LOD plot across a single chromosome (*see* **Notes 30** and **31**).

```
>plot(scan,chr=1)
```

3. Analyze the effect plots at the peak LOD of your QTL to assess mode of inheritance (*see* Fig. 1, **Notes 32** and **33**). The *effect* command will output the phenotypic mean and standard error

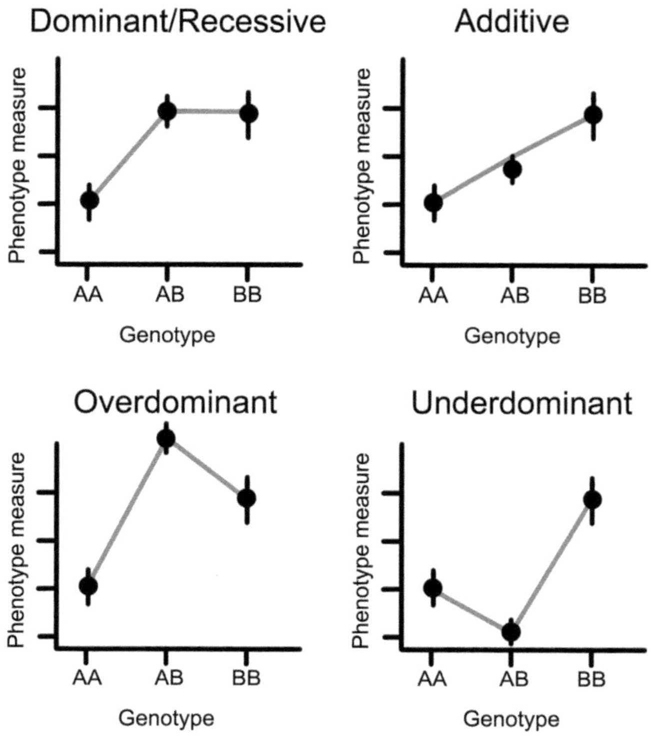

Fig. 1 Example effect plots demonstrating different modes of inheritance

values (*see* **Note 34**). The below command is for the locus on chromosome 1 at 5 cM.

```
>effect<-effectplot(augdata,pheno.col=2,mname1= find.
marker(augdata,chr=1,pos=5))
>effect
```

4. Calculate the additive, dominance effects, and heritability for the peak marker for each QTL locus (*see* **Note 34**), replacing AA, AB, and BB in the below formulas with the mean values outputted from the *effect* command.

```
>effect
>add<-((AA-BB)/2)
>add
>dom<-(AB -(AA+BB)/2)
>dom
>heritability<-(2*add^2+dom^2)/(2*add^2+dom^2+4*1)
>heritability
```

5. Calculate the phenotypic variance explained (PVE) by the QTL (*see* **Note 35**), replacing n in the below formula with the

number of individuals included in the analysis (*see* **Note 9**) and LOD in the below with the outputted LOD score for that locus.

```
>PVE<-(1-(10^-((2/n)*(LOD))))
>PVE
```

3.6 From QTL to Candidate Genes and Causative Loci

1. QTL loci are commonly large genetic intervals (*see* **Note 31**), and considerable additional work is necessary to determine candidate genes and causative genetic changes. Often, candidate genes within a QTL interval are chosen based on bioinformatic analysis, literature searches, and other genetic information such as population genetics. Defining causal mutations is particularly difficult and requires a great deal of follow-up work including experiments such as gene expression analysis, embryonic manipulations, characterization of loss-of-function phenotypes, and gene editing (e.g., using the CRISPR-Cas9 system).

4 Notes

1. Data should be in .csv format. This format can easily be produced from a spreadsheet in OpenOffice or Microsoft Excel. The first column should be an individual ID, followed by columns as needed with data on one phenotype each. The remaining columns are genotype data, one per marker. The first row is a header, with a description of the phenotype or the marker name. The second and third rows are reserved for genetic map information, or are left blank if a genetic map is being estimated (*see* **Note 4**). This second contains chromosome number as a numeral and the third row contains centiMorgan (cM) position for all markers; the second and third rows must be empty for the ID and phenotype columns.
2. A .csv formatted mapping file in the appropriate format is supplied at https://github.com/kpowder/MiMB_QTL. This file is highly modified partial data set from [41]. When including all individuals and markers as given and building a model with markers 5 and 119 as cofactors, a significant QTL is present on chr 1. This QTL has an additive effect, LOD = 4.29, peak at marker 5, and a 95% CI from marker 1 to marker 22 (0–18.2 cM). This QTL explains 7.66% of the total phenotypic variation for this trait (i.e., PVE=0.0766).
3. Additional genotype formats are acceptable, for example, "A," "B," and "C" for a microsatellite with three different lengths. If using a different set of strings for genotypes or missing data, this can be adjusted in the command in Subheading 3.1, **step 4**.

4. If the genetic map has not been assembled yet, it can be estimated at this time by including *estimate.map=TRUE* in the *read.cross* command, such as below. Be warned that including a map estimation will make this command take a while to process.

```
>data<-read.cross(format="csv", file=file.choose(),
genotypes=c("AA", "AB", "BB"), na.strings=c("NA"),
estimate.map=TRUE, convertXdata=TRUE)
```

5. The number of phenotypes listed in by the *summary(data)* command will include any non-genotype columns. For instance, in the sample data file, there are three "phenotypes" of ID, the phenotype data, and sex.
6. Full genotype data can be attained below, including genotype frequencies at each marker and a *p* value for tests of 1:2:1 Mendelian proportions expected in an intercross.

```
>geno.table(data)
```

7. In this visual, AA genotypes are represented by red, BB by green, AB by blue, and NA by white colors. Rows are individuals and columns are markers.
8. In this visual, black marks indicate missing genotypic information. Rows are individuals and columns are markers.
9. Missing genotypes would need to be excluded from the statistical model, reducing power. Using the *mqmaugment* function statistically predicts missing genotypes based on neighboring markers and recombination rates [11]. Genotypes are modeled and all possible genotypes are created as new individuals with weighted probabilities. When visualizing the augmented data with the below, you will note that there are no longer white (missing) genotypes and the number of individuals increases based on the weighted genotypes. Note that any time you need the number of individuals (e.g., calculating PVE in Subheading 3.5, **step 5**), you should use the original number of individuals included rather than the number of individuals after augmentation.

```
>geno.image(augdata)
```

10. The program counts column A as column 1 when counting columns.
11. I generally use *model = "dominant"* as this will assess alleles with both dominant and additive effects. The default option *model = "additive"* will only assess for additive effects.

12. This summary command will list only the top LOD scores across the genome. Additional useful commands are below.

```
>scan #displays LOD scores at every marker
>write.csv(scan,file="scan_LODs.csv") #outputs all LOD scores
as a csv file
```

13. Use of cofactors in the model eliminates phenotypic variation due to other QTL and more accurately estimates the effect of a QTL.
14. As a rule of thumb, any locus with a LOD >2.5–3 is a putative QTL. Once the full model is built, a LOD threshold will be empirically determined using permutations (*see* Subheading 3.4, **step 1**).
15. If there are no obvious LOD "peaks" of >2.5, you may want to run an autoscan. The below commands will randomly pick 250 loci across the genome, accounting for marker density, that can be used as an initial set of cofactors to build the model.

```
>cofactors <- mqmautocofactors(augdata,250)
```

16. In order to treat sex as a cofactor, include the sex-determining locus as a cofactor. If other fixed cofactors such as family structure, environment, or diet are necessary to include, you may need to build models and conduct backward elimination manually using the *scanone* function (*see* [9]).
17. Running >*cofactorslist* once you have entered all your putative QTL will give you a list of marker numbers to ensure you've got them all.
18. While loci are commonly outputted as a chromosome number and chromosomal position, the software also assigns each marker a number. I find it helpful to keep a scratch piece of paper or file with both of these pieces of information. The below will display the marker number based on chromosomal location (here, at 5 cM on chromosome 1).

```
>find.marker(augdata,chr=1,pos=5)
```

19. Running >*cofactors* after this will give you a matrix of all markers in your data set. Those included as a cofactor will be a "1" and those not included are "0". This matrix can be overwhelming, but a count of "1" entries is quick check that you have the expected number of cofactors included.
20. If you know the marker number (*see* **Note 18**), you can quickly adjust this matrix without adjusting the list of cofactors (i.e.,

not going back to Subheadings 3.3, **step 4** or **5**). In the example below, marker 34 is switched to be included, while marker 78 is switched to be excluded.

```
>cofactors[34]<-1
>cofactors[78]<-0
```

21. Use the below to visualize how your cofactors are spread across your chromosomes.

```
>mqmplot.cofactors(augdata, cofactors, justdots=T)
```

22. This analysis conducts an unsupervised backward elimination to verify cofactors, meaning a cofactor is removed and the analysis is recalculated. I generally remove the eliminated cofactors from the next iteration of the model as too many cofactors decreases power [82]. Note that cofactors eliminated from one model may not be eliminated from all models.
23. Again, a piece of scratch paper can come in handy as you run through different iterations of the models. I generally record the locus chromosome and position information and marker number of cofactors in the model, as well as which of these were eliminated.
24. I generally consider the model "stable" when (1) all putative QTL peaks are included as cofactors and retained in the model or (2) any putative QTL peak not included in the model is eliminated when added.
25. The total number of cofactors in the model should not exceed two times the square root of the number of individuals [94].
26. Generally, loci that pass the 5% threshold are significant QTL, while those that are above the 10% are considered suggestive.
27. Bonferroni correction is not appropriate to adjust for multiple testing given that markers are potentially linked. Rather, QTL mapping uses randomized permutations to empirically establish a significance threshold [95]. This permutation also accounts for variation in the data set such as number of individuals, number of markers, pattern of missing data, and the phenotypic variation. Most commonly, 1000 permutations are conducted.
28. Remember that a LOD of >2.5–3 was used as a rule of thumb for putative QTL when building the model. If the permuted significance level is less than this value, you should look at the effect plots for the QTL that now would be considered significant and adjust your model if needed. If you adjust the model, you will need to re-rerun the permutation.

29. Another option instead of calculating peak interval with Bayesian analysis is to determine where there is a drop of 1.5 in the LOD score from the peak. However, the 1.5-lod support interval can vary greatly [96]. Replace the "X" in the below with your chromosome number.

```
>lodint(scan,X,qtl.index=1,drop=1.5,lodcolumn=1,
expandtomarkers=T)
```

30. Plots of LOD score can be used to help clarify if there may be two closely linked QTL. If there is a dip in the LOD score, this can be suggestive of multiple peaks in the same region rather than a single, large peak, though additional work is necessary to confirm this. Looking at effect plots across the region may also help distinguish if effects are consistent across the region or have variation that may suggest multiple peaks.
31. The level of resolution for a QTL analysis is based on density of markers and the amount of recombination; for many F_2 QTL analyses the limit of resolution is in the range of 3–20 cM [10, 97]. Thus, it may not be possible to distinguish between a single QTL locus and two QTL that are tightly linked. Finer mapping of the QTL interval can be conducted in a later generation after additional rounds of inbreeding (advanced intercross lines).
32. Two-way effect plots can be used to assess nonadditive effects between two specific genetic loci (also called gene interactions or epistasis) [98–100]. The below command is for the loci on chromosome 1 at 5 cM and chromosome 3 at 25 cM.

```
>effect2<-effectplot(data,pheno.col=2,mname1= find.marker
(augdata, chr=1, pos=5), mname2= find.marker(augdata,
chr=3, pos=25))
>effect2
```

33. It is possible to conduct genome-wide scans for epistatic interactions, in which the model is fitted to every possible pairwise combination of loci [9, 98, 99]. This has a variety of challenges including increased sample sizes due to partitioning of genotype groups, corrections for multiple testing, and increased computational demands that often require parallel computing [98, 99].
34. It is common practice to report the additive, dominant, and heritability values (*see* Subheading 3.5, **step 4**) for the peak marker of each QTL in tables in publications, as well as mean phenotypic values from effect plot (*see* Subheading 3.5, **step 3**). Additionally, the LOD score (*see* Subheading 3.3, **step 2** and **Note 12**), confidence interval (*see* Subheading 3.4, **step 2**),

and PVE (*see* Subheading 3.5, **step 5**) should be reported in publications.

35. Small sample sizes and thus decreased power can lead to an overestimation of the phenotypic variance explained (PVE), called the Beavis effect [101, 102].

References

1. Alberch P (1991) From genes to phenotype: dynamical systems and evolvability. Genetica 84(1):5–11
2. Wagner GP, Altenberg L (1996) Perspective: complex adaptations and the evolution of evolvability. Evolution 50(3):967–976. https://doi.org/10.1111/j.1558-5646.1996.tb02339.x
3. Castle WE (1921) An improved method of estimating the number of genetic factors concerned in cases of blending inheritance. Science 54(1393):223. https://doi.org/10.1126/science.54.1393.223
4. Fisher RA (1918) The correlation between relatives on the supposition of Mendelian inheritance. Trans R Soc Edinburgh 52:399–433
5. Haldane JBS (1932) The causes of evolution. Harper and Brothers, London
6. Wright S (1921) Systems of mating. I. the biometric relations between parent and offspring. Genetics 6(2):111–123
7. Wright S (1968) Evolution and genetics of populations, vol 1. University of Chicago Press, Chicago
8. Lynch M, Walsh B (1998) Genetics and analysis of quantitative traits. Sinauer, Sunderland, MA
9. Broman KW, Sen S (2009) A guide to QTL mapping with R/qtl. Springer, New York
10. Falconer DS, Mackay TFC (2009) Introduction to quantitative genetics, 4th edn. Pearson, London
11. Arends D, Prins P, Jansen RC, Broman KW (2010) R/qtl: high-throughput multiple QTL mapping. Bioinformatics 26 (23):2990–2992. https://doi.org/10.1093/bioinformatics/btq565
12. Rifkin SA (2012) Quantitative trait loci (QTL): methods and protocols. Methods in molecular biology. Humana Press, Totowa, NJ
13. Doerge RW (2002) Mapping and analysis of quantitative trait loci in experimental populations. Nat Rev Genet 3(1):43–52. https://doi.org/10.1038/nrg703
14. Mackay TF, Stone EA, Ayroles JF (2009) The genetics of quantitative traits: challenges and prospects. Nat Rev Genet 10(8):565–577. https://doi.org/10.1038/nrg2612
15. Mackay TF, Fry JD (1996) Polygenic mutation in *Drosophila melanogaster*: genetic interactions between selection lines and candidate quantitative trait loci. Genetics 144 (2):671–688
16. Mackay TF (1996) The nature of quantitative genetic variation revisited: lessons from *Drosophila* bristles. BioEssays 18(2):113–121. https://doi.org/10.1002/bies.950180207
17. Keightley PD, Hardge T, May L, Bulfield G (1996) A genetic map of quantitative trait loci for body weight in the mouse. Genetics 142 (1):227–235
18. Johnson TE, DeFries JC, Markel PD (1992) Mapping quantitative trait loci for behavioral traits in the mouse. Behav Genet 22 (6):635–653
19. Cheverud JM, Routman EJ, Duarte FA, van Swinderen B, Cothran K, Perel C (1996) Quantitative trait loci for murine growth. Genetics 142(4):1305–1319
20. Diers BW, Keim P, Fehr WR, Shoemaker RC (1992) RFLP analysis of soybean seed protein and oil content. Theor Appl Genet 83 (5):608–612. https://doi.org/10.1007/BF00226905
21. Pe ME, Gianfranceschi L, Taramino G, Tarchini R, Angelini P, Dani M, Binelli G (1993) Mapping quantitative trait loci (QTLs) for resistance to *Gibberella zeae* infection in maize. Mol Gen Genet 241 (1-2):11–16
22. Laurie DA, Pratchett N, Snape JW, Bezant JH (1995) RFLP mapping of five major genes and eight quantitative trait loci controlling flowering time in a winter x spring barley (*Hordeum vulgare* L.) cross. Genome 38 (3):575–585
23. Veldboom LR, Lee M (1994) Molecular-marker-facilitated studies of morphological traits in maize. II: determination of QTLs for grain yield and yield components. Theor

Appl Genet 89(4):451–458. https://doi.org/10.1007/BF00225380
24. Edwards MD, Stuber CW, Wendel JF (1987) Molecular-marker-facilitated investigations of quantitative-trait loci in maize. I. Numbers, genomic distribution and types of gene action. Genetics 116(1):113–125
25. Sucena E, Stern DL (2000) Divergence of larval morphology between *Drosophila sechellia* and its sibling species caused by cis-regulatory evolution of ovo/shaven-baby. Proc Natl Acad Sci U S A 97(9):4530–4534
26. Liu J, Mercer JM, Stam LF, Gibson GC, Zeng ZB, Laurie CC (1996) Genetic analysis of a morphological shape difference in the male genitalia of *Drosophila simulans* and *D. mauritiana*. Genetics 142(4):1129–1145
27. True JR, Liu J, Stam LF, Zeng ZB, Laurie CC (1997) Quantitative genetic analysis of divergence in male secondary sexual traits between *Drosophila simulans* and *Drosophila mauritiana*. Evolution 51(3):816–832. https://doi.org/10.1111/j.1558-5646.1997.tb03664.x
28. Doebley J (2004) The genetics of maize evolution. Annu Rev Genet 38:37–59. https://doi.org/10.1146/annurev.genet.38.072902.092425
29. Kodama M, Hard JJ, Naish KA (2018) Mapping of quantitative trait loci for temporal growth and age at maturity in coho salmon: evidence for genotype-by-sex interactions. Mar Genomics 38:33–44. https://doi.org/10.1016/j.margen.2017.07.004
30. Fu B, Liu H, Yu X, Tong J (2016) A high-density genetic map and growth related QTL mapping in bighead carp (*Hypophthalmichthys nobilis*). Sci Rep 6:28679. https://doi.org/10.1038/srep28679
31. Wringe BF, Devlin RH, Ferguson MM, Moghadam HK, Sakhrani D, Danzmann RG (2010) Growth-related quantitative trait loci in domestic and wild rainbow trout (*Oncorhynchus mykiss*). BMC Genet 11:63. https://doi.org/10.1186/1471-2156-11-63
32. Liu H, Fu B, Pang M, Feng X, Yu X, Tong J (2017) A high-density genetic linkage map and QTL fine mapping for body weight in crucian carp (*Carassius auratus*) using 2b-RAD sequencing. G3 (Bethesda) 7(8):2473–2487. https://doi.org/10.1534/g3.117.041376
33. Lin G, Chua E, Orban L, Yue GH (2016) Mapping QTL for sex and growth traits in salt-tolerant tilapia (*Oreochromis* spp. X *O. mossambicus*). PLoS One 11(11):e0166723. https://doi.org/10.1371/journal.pone.0166723
34. Wang L, Wan ZY, Bai B, Huang SQ, Chua E, Lee M, Pang HY, Wen YF, Liu P, Liu F, Sun F, Lin G, Ye BQ, Yue GH (2015) Construction of a high-density linkage map and fine mapping of QTL for growth in Asian seabass. Sci Rep 5:16358. https://doi.org/10.1038/srep16358
35. Miller CT, Glazer AM, Summers BR, Blackman BK, Norman AR, Shapiro MD, Cole BL, Peichel CL, Schluter D, Kingsley DM (2014) Modular skeletal evolution in sticklebacks is controlled by additive and clustered quantitative trait Loci. Genetics 197(1):405–420. https://doi.org/10.1534/genetics.114.162420
36. Albertson RC, Streelman JT, Kocher TD, Yelick PC (2005) Integration and evolution of the cichlid mandible: the molecular basis of alternate feeding strategies. Proc Natl Acad Sci U S A 102(45):16287–16292. https://doi.org/10.1073/pnas.0506649102
37. Albertson RC, Streelman JT, Kocher TD (2003) Directional selection has shaped the oral jaws of Lake Malawi cichlid fishes. Proc Natl Acad Sci U S A 100(9):5252–5257. https://doi.org/10.1073/pnas.0930235100
38. Hulsey CD, Machado-Schiaffino G, Keicher L, Ellis-Soto D, Henning F, Meyer A (2017) The integrated genomic architecture and evolution of dental divergence in East African cichlid fishes (*Haplochromis chilotes* x *H. nyererei*). G3 (Bethesda) 7(9):3195–3202. https://doi.org/10.1534/g3.117.300083
39. Streelman JT, Albertson RC (2006) Evolution of novelty in the cichlid dentition. J Exp Zool B Mol Dev Evol 306(3):216–226. https://doi.org/10.1002/jez.b.21101
40. Peichel CL, Nereng KS, Ohgi KA, Cole BL, Colosimo PF, Buerkle CA, Schluter D, Kingsley DM (2001) The genetic architecture of divergence between threespine stickleback species. Nature 414(6866):901–905. https://doi.org/10.1038/414901a
41. Albertson RC, Kawasaki KC, Tetrault ER, Powder KE (2018) Genetic analyses in Lake Malawi cichlids identify new roles for Fgf signaling in scale shape variation. Commun Biol 1:55. https://doi.org/10.1038/s42003-018-0060-4
42. Navon D, Olearczyk N, Albertson RC (2017) Genetic and developmental basis for fin shape variation in African cichlid fishes. Mol Ecol 26(1):291–303. https://doi.org/10.1111/mec.13905
43. O'Quin KE, Yoshizawa M, Doshi P, Jeffery WR (2013) Quantitative genetic analysis of

retinal degeneration in the blind cavefish *Astyanax mexicanus*. PLoS One 8(2):e57281. https://doi.org/10.1371/journal.pone.0057281
44. Liu J, Shikano T, Leinonen T, Cano JM, Li MH, Merila J (2014) Identification of major and minor QTL for ecologically important morphological traits in three-spined sticklebacks (*Gasterosteus aculeatus*). G3 (Bethesda) 4(4):595–604. https://doi.org/10.1534/g3.114.010389
45. Cresko WA, Amores A, Wilson C, Murphy J, Currey M, Phillips P, Bell MA, Kimmel CB, Postlethwait JH (2004) Parallel genetic basis for repeated evolution of armor loss in Alaskan threespine stickleback populations. Proc Natl Acad Sci U S A 101(16):6050–6055. https://doi.org/10.1073/pnas.0308479101
46. Colosimo PF, Peichel CL, Nereng K, Blackman BK, Shapiro MD, Schluter D, Kingsley DM (2004) The genetic architecture of parallel armor plate reduction in threespine sticklebacks. PLoS Biol 2(5):E109. https://doi.org/10.1371/journal.pbio.0020109
47. Shapiro MD, Marks ME, Peichel CL, Blackman BK, Nereng KS, Jonsson B, Schluter D, Kingsley DM (2004) Genetic and developmental basis of evolutionary pelvic reduction in threespine sticklebacks. Nature 428 (6984):717–723. https://doi.org/10.1038/nature02415
48. Klingenberg CP (2010) Evolution and development of shape: integrating quantitative approaches. Nat Rev Genet 11(9):623–635. https://doi.org/10.1038/nrg2829
49. Mitteroecker P, Gunz P (2009) Advances in geometric morphometrics. Evol Biol 36:235–247
50. Li Z, Guo B, Yang J, Herczeg G, Gonda A, Balazs G, Shikano T, Calboli FC, Merila J (2017) Deciphering the genomic architecture of the stickleback brain with a novel multilocus gene-mapping approach. Mol Ecol 26 (6):1557–1575. https://doi.org/10.1111/mec.14005
51. Stewart TA, Albertson RC (2010) Evolution of a unique predatory feeding apparatus: functional anatomy, development and a genetic locus for jaw laterality in Lake Tanganyika scale-eating cichlids. BMC Biol 8:8. https://doi.org/10.1186/1741-7007-8-8
52. Franchini P, Fruciano C, Spreitzer ML, Jones JC, Elmer KR, Henning F, Meyer A (2014) Genomic architecture of ecologically divergent body shape in a pair of sympatric crater lake cichlid fishes. Mol Ecol 23 (7):1828–1845. https://doi.org/10.1111/mec.12590
53. Fruciano C, Franchini P, Kovacova V, Elmer KR, Henning F, Meyer A (2016) Genetic linkage of distinct adaptive traits in sympatrically speciating crater lake cichlid fish. Nat Commun 7:12736. https://doi.org/10.1038/ncomms12736
54. Parsons KJ, Wang J, Anderson G, Albertson RC (2015) Nested levels of adaptive divergence: the genetic basis of craniofacial divergence and ecological sexual dimorphism. G3 (Bethesda) 5(8):1613–1624. https://doi.org/10.1534/g3.115.018226
55. Streelman JT, Albertson RC, Kocher TD (2003) Genome mapping of the orange blotch colour pattern in cichlid fishes. Mol Ecol 12(9):2465–2471
56. Albertson RC, Powder KE, Hu Y, Coyle KP, Roberts RB, Parsons KJ (2014) Genetic basis of continuous variation in the levels and modular inheritance of pigmentation in cichlid fishes. Mol Ecol 23(21):5135–5150. https://doi.org/10.1111/mec.12900
57. Gross JB, Borowsky R, Tabin CJ (2009) A novel role for Mc1r in the parallel evolution of depigmentation in independent populations of the cavefish *Astyanax mexicanus*. PLoS Genet 5(1):e1000326. https://doi.org/10.1371/journal.pgen.1000326
58. Yong L, Peichel CL, McKinnon JS (2015) Genetic architecture of conspicuous red ornaments in female threespine stickleback. G3 (Bethesda) 6(3):579–588. https://doi.org/10.1534/g3.115.024505
59. Tsuboko S, Kimura T, Shinya M, Suehiro Y, Okuyama T, Shimada A, Takeda H, Naruse K, Kubo T, Takeuchi H (2014) Genetic control of startle behavior in medaka fish. PLoS One 9 (11):e112527. https://doi.org/10.1371/journal.pone.0112527
60. Greenwood AK, Ardekani R, McCann SR, Dubin ME, Sullivan A, Bensussen S, Tavare S, Peichel CL (2015) Genetic mapping of natural variation in schooling tendency in the threespine stickleback. G3 (Bethesda) 5 (5):761–769. https://doi.org/10.1534/g3.114.016519
61. Wright D, Butlin RK, Carlborg O (2006) Epistatic regulation of behavioural and morphological traits in the zebrafish (*Danio rerio*). Behav Genet 36(6):914–922. https://doi.org/10.1007/s10519-006-9080-9
62. Wright D, Nakamichi R, Krause J, Butlin RK (2006) QTL analysis of behavioral and morphological differentiation between wild and laboratory zebrafish (*Danio rerio*). Behav

Genet 36(2):271–284. https://doi.org/10.1007/s10519-005-9029-4

63. Waits ER, Nebert DW (2011) Genetic architecture of susceptibility to PCB126-induced developmental cardiotoxicity in zebrafish. Toxicol Sci 122(2):466–475. https://doi.org/10.1093/toxsci/kfr136
64. Nacci D, Proestou D, Champlin D, Martinson J, Waits ER (2016) Genetic basis for rapidly evolved tolerance in the wild: adaptation to toxic pollutants by an estuarine fish species. Mol Ecol 25(21):5467–5482. https://doi.org/10.1111/mec.13848
65. Palaiokostas C, Cariou S, Bestin A, Bruant JS, Haffray P, Morin T, Cabon J, Allal F, Vandeputte M, Houston RD (2018) Genome-wide association and genomic prediction of resistance to viral nervous necrosis in European sea bass (*Dicentrarchus labrax*) using RAD sequencing. Genet Sel Evol 50 (1):30. https://doi.org/10.1186/s12711-018-0401-2
66. Wang L, Liu P, Huang S, Ye B, Chua E, Wan ZY, Yue GH (2017) Genome-Wide Association Study identifies loci associated with resistance to viral nervous necrosis disease in Asian Seabass. Mar Biotechnol (NY) 19 (3):255–265. https://doi.org/10.1007/s10126-017-9747-7
67. Wang L, Bai B, Huang S, Liu P, Wan ZY, Ye B, Wu J, Yue GH (2017) QTL mapping for resistance to Iridovirus in Asian Seabass using genotyping-by-sequencing. Mar Biotechnol (NY) 19(5):517–527. https://doi.org/10.1007/s10126-017-9770-8
68. Liu S, Vallejo RL, Gao G, Palti Y, Weber GM, Hernandez A, Rexroad CE 3rd (2015) Identification of single-nucleotide polymorphism markers associated with cortisol response to crowding in rainbow trout. Mar Biotechnol (NY) 17(3):328–337. https://doi.org/10.1007/s10126-015-9621-4
69. Kusakabe M, Ishikawa A, Ravinet M, Yoshida K, Makino T, Toyoda A, Fujiyama A, Kitano J (2017) Genetic basis for variation in salinity tolerance between stickleback ecotypes. Mol Ecol 26 (1):304–319. https://doi.org/10.1111/mec.13875
70. Haidle L, Janssen JE, Gharbi K, Moghadam HK, Ferguson MM, Danzmann RG (2008) Determination of quantitative trait loci (QTL) for early maturation in rainbow trout (*Oncorhynchus mykiss*). Mar Biotechnol (NY) 10(5):579–592. https://doi.org/10.1007/s10126-008-9098-5
71. Wan SM, Liu H, Zhao BW, Nie CH, Wang WM, Gao ZX (2017) Construction of a high-density linkage map and fine mapping of QTLs for growth and gonad related traits in blunt snout bream. Sci Rep 7:46509. https://doi.org/10.1038/srep46509
72. Kliebenstein D (2009) Quantitative genomics: analyzing intraspecific variation using global gene expression polymorphisms or eQTLs. Annu Rev Plant Biol 60:93–114. https://doi.org/10.1146/annurev.arplant.043008.092114
73. Brown KH, Dobrinski KP, Lee AS, Gokcumen O, Mills RE, Shi X, Chong WW, Chen JY, Yoo P, David S, Peterson SM, Raj T, Choy KW, Stranger BE, Williamson RE, Zon LI, Freeman JL, Lee C (2012) Extensive genetic diversity and substructuring among zebrafish strains revealed through copy number variant analysis. Proc Natl Acad Sci U S A 109(2):529–534. https://doi.org/10.1073/pnas.1112163109
74. Uusi-Heikkila S, Savilammi T, Leder E, Arlinghaus R, Primmer CR (2017) Rapid, broad-scale gene expression evolution in experimentally harvested fish populations. Mol Ecol 26(15):3954–3967. https://doi.org/10.1111/mec.14179
75. Ishikawa A, Kusakabe M, Yoshida K, Ravinet M, Makino T, Toyoda A, Fujiyama A, Kitano J (2017) Different contributions of local- and distant-regulatory changes to transcriptome divergence between stickleback ecotypes. Evolution 71 (3):565–581. https://doi.org/10.1111/evo.13175
76. Pritchard VL, Viitaniemi HM, McCairns RJ, Merila J, Nikinmaa M, Primmer CR, Leder EH (2017) Regulatory architecture of gene expression variation in the threespine stickleback *Gasterosteus aculeatus*. G3 (Bethesda) 7 (1):165–178. https://doi.org/10.1534/g3.116.033241
77. Pavlicev M, Cheverud JM, Wagner GP (2011) Evolution of adaptive phenotypic variation patterns by direct selection for evolvability. Proc Biol Sci 278(1713):1903–1912. https://doi.org/10.1098/rspb.2010.2113
78. Hu Y, Parsons KJ, Albertson RC (2014) Evolvability of the cichlid jaw: new tools provide insights into the genetic basis of phenotypic integration. Evol Biol 41(1):145–153
79. Parsons KJ, Marquez E, Albertson RC (2012) Constraint and opportunity: the genetic basis and evolution of modularity in the cichlid mandible. Am Nat 179(1):64–78. https://doi.org/10.1086/663200
80. Jansen RC (1993) Interval mapping of multiple quantitative trait loci. Genetics 135 (1):205–211

81. Jansen RC (1994) Controlling the type I and type II errors in mapping quantitative trait loci. Genetics 138(3):871–881
82. Zeng ZB (1994) Precision mapping of quantitative trait loci. Genetics 136(4):1457–1468
83. Visscher PM, Hill WG, Wray NR (2008) Heritability in the genomics era—concepts and misconceptions. Nat Rev Genet 9 (4):255–266. https://doi.org/10.1038/nrg2322
84. Otto SP, Jones CD (2000) Detecting the undetected: estimating the total number of loci underlying a quantitative trait. Genetics 156(4):2093–2107
85. Sen S, Satagopan JM, Broman KW, Churchill GA (2007) R/qtlDesign: inbred line cross experimental design. Mamm Genome 18 (2):87–93. https://doi.org/10.1007/s00335-006-0090-y
86. Davey JW, Hohenlohe PA, Etter PD, Boone JQ, Catchen JM, Blaxter ML (2011) Genome-wide genetic marker discovery and genotyping using next-generation sequencing. Nat Rev Genet 12(7):499–510. https://doi.org/10.1038/nrg3012
87. Jamann TM, Balint-Kurti PJ, Holland JB (2015) QTL mapping using high-throughput sequencing. Methods Mol Biol 1284:257–285. https://doi.org/10.1007/978-1-4939-2444-8_13
88. Baird NA, Etter PD, Atwood TS, Currey MC, Shiver AL, Lewis ZA, Selker EU, Cresko WA, Johnson EA (2008) Rapid SNP discovery and genetic mapping using sequenced RAD markers. PLoS One 3(10):e3376. https://doi.org/10.1371/journal.pone.0003376
89. Chutimanitsakun Y, Nipper RW, Cuesta-Marcos A, Cistue L, Corey A, Filichkina T, Johnson EA, Hayes PM (2011) Construction and application for QTL analysis of a Restriction Site Associated DNA (RAD) linkage map in barley. BMC Genomics 12:4. https://doi.org/10.1186/1471-2164-12-4
90. Sonah H, Bastien M, Iquira E, Tardivel A, Legare G, Boyle B, Normandeau E, Laroche J, Larose S, Jean M, Belzile F (2013) An improved genotyping by sequencing (GBS) approach offering increased versatility and efficiency of SNP discovery and genotyping. PLoS One 8(1):e54603. https://doi.org/10.1371/journal.pone.0054603
91. Elshire RJ, Glaubitz JC, Sun Q, Poland JA, Kawamoto K, Buckler ES, Mitchell SE (2011) A robust, simple genotyping-by-sequencing (GBS) approach for high diversity species. PLoS One 6(5):e19379. https://doi.org/10.1371/journal.pone.0019379
92. Schlotterer C, Tobler R, Kofler R, Nolte V (2014) Sequencing pools of individuals—mining genome-wide polymorphism data without big funding. Nat Rev Genet 15 (11):749–763. https://doi.org/10.1038/nrg3803
93. Van Ooijen J (2006) JoinMap 4. Software for the calculation of genetic linkage maps in experimental populations. Kayazama BV, Wageningen
94. Jansen RC, Stam P (1994) High resolution of quantitative traits into multiple loci via interval mapping. Genetics 136(4):1447–1455
95. Churchill GA, Doerge RW (1994) Empirical threshold values for quantitative trait mapping. Genetics 138(3):963–971
96. Papachristou C, Lin S (2006) A comparison of methods for intermediate fine mapping. Genet Epidemiol 30(8):677–689. https://doi.org/10.1002/gepi.20179
97. Mackay TF (2001) Quantitative trait loci in *Drosophila*. Nat Rev Genet 2(1):11–20. https://doi.org/10.1038/35047544
98. Carlborg O, Haley CS (2004) Epistasis: too often neglected in complex trait studies? Nat Rev Genet 5(8):618–625. https://doi.org/10.1038/nrg1407
99. Mackay TF (2014) Epistasis and quantitative traits: using model organisms to study gene-gene interactions. Nat Rev Genet 15 (1):22–33. https://doi.org/10.1038/nrg3627
100. Phillips PC (2008) Epistasis—the essential role of gene interactions in the structure and evolution of genetic systems. Nat Rev Genet 9 (11):855–867. https://doi.org/10.1038/nrg2452
101. Beavis WD (1998) QTL analysis: power, precision, and accuracy. In: Molecular dissection of complex traits. CRC Press, Boca Raton
102. Xu S (2003) Theoretical basis of the Beavis effect. Genetics 165(4):2259–2268

Chapter 16

Expression Quantitative Trait Loci Analysis in Multiple Tissues

Gen Li

Abstract

Expression quantitative trait loci (eQTL) analysis identifies genetic variants that regulate the expression level of a gene. The genetic regulation may persist or vary in different tissues. When data are available on multiple tissues, it is often desired to borrow information across tissues and conduct an integrative analysis. Here we describe a multi-tissue eQTL analysis procedure, which improves the identification of different types of eQTL and facilitates the assessment of tissue specificity.

Key words eQTL analysis, Multiple tissues, Data integration, Tissue specificity, Hypothesis testing

1 Introduction

Expression quantitative trait loci (eQTL) analysis has drawn significant attention over the past few years due to its success in advancing our knowledge of genetic pathways and networks underlying complex human traits [1]. Standard eQTL analysis directly tests the association between a single nucleotide polymorphism (SNP) and the expression level of a gene. There are many powerful statistical software packages for such analysis [2–4]. To date, numerous eQTL have been found to be associated with important traits such as complex diseases [5].

Recently, large collaborative efforts such as the Genotype-Tissue Expression (GTEx) project [6, 7] collect genetic data from multiple tissues, enabling multi-tissue eQTL analysis. On one hand, the association between a gene and a SNP may vary in different tissues. A multi-tissue study has the potential to elucidate the genetic difference between tissues by identifying tissue-specific eQTL [6]. On the other hand, many eQTL are shared across tissues [6]. Namely, if a SNP regulates a gene in one tissue, it is highly likely to regulate the same gene in other tissues as well. Thus, one

Xinghua Mindy Shi (ed.), *eQTL Analysis: Methods and Protocols*, Methods in Molecular Biology, vol. 2082,
https://doi.org/10.1007/978-1-0716-0026-9_16,

may borrow strength across tissues to improve findings in a single tissue.

There have been some recent developments on multi-tissue eQTL analysis methods [8–12]. Most existing methods are computationally intensive and do not scale well to large data. Here we elaborate a scalable algorithm for multi-tissue eQTL analysis [9]. The method takes standard single-tissue eQTL summary statistics as input and builds upon a hierarchical Bayesian model to make inferences. The model fitting procedure follows an empirical Bayes approach, which is computationally efficient. It can be easily implemented on a standard desktop. The model parameters estimated from data have critical biological interpretations, and inform eQTL configurations in multiple tissues. We will carefully describe the analysis procedure step by step, from raw data preprocessing to final results presentation. We will demonstrate how to conduct hypothesis testing to identify different types of eQTL. More technical details of the method can be found in [8, 9].

2 Materials

We will focus on cis-eQTL analysis where a SNP is located near the transcription starting site (TSS) of a gene. In human studies, the cis window is usually set to be 100 kb or 1 Mb. The number of gene-SNP pairs is on the scale of several million. Typically, the analysis described below can be conducted on a standard desktop computer with at least 8 Gb of RAM within 24 h.

2.1 Raw Data

1. Genotype data. Data may be obtained from whole genome sequencing or SNP array methods. The data set should be curated in a matrix format where each row corresponds to a SNP and each column corresponds to a sample. Each entry represents the number of minor allele variants of a SNP in one sample. Missing data should be imputed.
2. Gene expression data. Curate the data set for each tissue in a matrix format where each row corresponds to a gene and each column corresponds to a sample. Different tissues may or may not have overlapping samples (*see* **Note 1**). Missing data should be imputed.
3. Covariate data. Commonly used covariates include sex, age, sequencing platform, top principal components of SNP data and/or gene expression data, etc. The covariates should be curated in a similar format to the gene expression data for each tissue—rows correspond to different covariates and columns correspond to different samples. The samples should be matched with those in the corresponding gene expression data matrix.

4. Gene location file. The file should have location information for each gene considered in the gene expression data, including chromosome number and starting and ending loci of the gene. Organize the data in a matrix format where each row is a gene and the columns correspond to the gene name (geneid), chromosome (chr), starting site (left), and ending site (right).
5. SNP location file. Similar to the gene location file, the SNP location file should contain the chromosome and location of each considered SNP.

2.2 Software and Packages

1. Install R software, which is freely available at https://www.r-project.org/. Install R version 3.2.1 or above to avoid incompatibility.
2. Install the R package "MatrixEQTL" [2] in R.
3. Install Matlab software from MathWork.
4. Download Matlab package "HT-eQTL" [9] from https://github.com/reagan0323/MT-eQTL/tree/master/HT-eQTL.

3 Methods

3.1 Data Preprocessing

1. Create a copy of the genotype data matrix for each tissue. Only keep the samples that are matched with the corresponding gene expression and covariates data.
2. Remove SNPs with low minor allele frequency (MAF) (*see* **Note 2**).
3. Remove genes with consistently low expression level (*see* **Note 3**).
4. Normalize gene expression data in each tissue by applying the inverse quantile normalization procedure to each gene. More specifically, for each gene, first order the expression levels across samples from low to high; then replace the kth (out of n) smallest expression level by $\Phi^{-1}(k/(n+1))$, where Φ is the standard normal cumulative distribution function. Ties will be replaced by the average rank. For example, if the 4th, 5th, and 6th smallest expression levels are the same, they will all be marked at 5th and inverse quantile normalized.
5. For each tissue, load the tailored genotype data, normalized gene expression data, covariates data, and gene/SNP location files into R (*see* **Note 4**).
6. Feed the data into the MatrixEQTL package to conduct single-tissue eQTL analysis and obtain summary statistics (i.e., t-statistics). Set the cis window size to be 1 kb (i.e., 1×10^5)

or 1 Mb (i.e., 1×10^6) and the p value threshold to be 1 in order to output summary statistics for all cis gene-SNP pairs.

7. Convert each summary statistic t to a correlation r in each tissue, using $r = \frac{t}{\sqrt{df+t^2}}$, where df is the degree of freedom (the number of samples minus the number of covariates minus two) in the corresponding tissue.
8. Further convert the correlations to z-statistics using Fisher transformation $z = \frac{1}{2}\sqrt{df-1}\ \ln\left(\frac{1+r}{1-r}\right)$.
9. Get the list of common gene-SNP pairs across all tissues.
10. Curate the obtained z-statistics into a matrix where each row corresponds to a gene-SNP pair in the common list and each column corresponds to a tissue. The z-statistics matrix is all we need for subsequent analyses.

3.2 Model Fitting

Essentially, one just needs to run the m function HT_pipeline.m in the Matlab package "HT-eQTL" from https://github.com/reagan0323/MT-eQTL/tree/master/HT-eQTL. The function output contains all estimated model parameters (*see* **Note 5**). Below, we provide a brief walk-through of the computational steps within the function.

1. For each pair of tissues, extract the two corresponding columns of the z-statistics matrix and fit an MT-eQTL model [8] using the EM_MT.m function in the Matlab package "HT-eQTL" (*see* **Note 6**).
2. Collect the Δ estimates from all pairs and assemble them into a single correlation matrix for all tissues (*see* **Note 7**).
3. Collect the Σ estimates from all pairs and assemble them into a single covariance matrix for all tissues (*see* **Note 8**).
4. Collect the probability mass function (pmf) estimates from all pairs and use a multi-probit model to convert it into a single pmf for all tissues.
5. Shrink small values in the obtained pmf to zero and renormalize the pmf (*see* **Note 9**).
6. Output the final model parameter estimates of Δ, Σ, and pmf for all tissues.

3.3 Statistical Inference

Once fit, the HT-eQTL model can be used to identify eQTL with different configurations (e.g., tissue-specific or tissue-common). Below, we particularly focus on the identification of gene-SNP pairs that are significantly associated (at a prefixed false discovery rate) in at least one of the studied tissues. A comprehensive simulation example is provided in HT_Sim_demo.m in the Matlab package "HT_eQTL." Additional examples can also be found in the manuscript [9].

1. Import the z-statistics matrix and the estimated model parameters (Δ, Σ, and pmf).
2. Calculate the posterior probability of each eQTL configuration for each gene-SNP pair using the MAposterior2_1.m function in the Matlab package "HT_eQTL." The output of the function is a posterior probability matrix with rows corresponding to gene-SNP pairs matched with the z-statistics matrix and columns corresponding to different eQTL configurations. In total, there are 2^K configurations where K is the number of studied tissues.
3. Extract the column corresponding to the null configuration (i.e., no eQTL in any tissue) from the posterior probability matrix. This column represents the local false discovery rates for all gene-SNP pairs under this hypothesis testing.
4. Sort the entries of the column in ascending order from smallest to largest.
5. Find the largest cutoff where the average of the entries below the cutoff is below the prefixed false discovery rate level (e.g., 0.05).
6. The gene-SNP pairs below the cutoff are identified as significantly associated in at least one of the studied tissues at the given false discovery rate.
7. If interested in identifying eQTL with other configurations, one just needs to calculate the posterior probabilities of complement configurations as the local false discovery rates and repeat **steps 4–6**.
8. One could also use the posterior probability matrix to determine the most probable eQTL configuration for each gene-SNP pair.

4 Notes

1. The presented method can deal with multi-tissue data with or without overlapping samples in different tissues. The model parameter Δ captures the correlation arising from overlapping samples in different tissues. If there is no overlap between tissues—namely, samples in different tissues come from different populations—one could fix Δ as an identify matrix without estimation. All subsequent analyses and inferences remain the same.
2. Per GTEx standard, genetic variants with MAF<1% are not considered.

3. Per GTEx standard, a gene will be selected if its reads per kilobase million (RPKM) value exceeds 0.1 in at least 20% of samples in a tissue.
4. We recommend using the *SlicedData* class in the R package "MatrixEQTL" to read and store large genetic files. The class is created for fast and memory-efficient manipulations of large datasets and compatible with other functions in the package.
5. The method is most suitable for a moderate number of tissues, say <20. If the number of tissues is too large, high computational cost and memory consumption may incur. One could conduct the analysis on different subsets of tissues.
6. If z-statistics are used, just set the degree of freedom for each tissue to be two in the input of the EM_MT.m function.
7. To assemble the K-by-K correlation matrix Δ (K being the number of studied tissues), take the off-diagonal value of each estimated correlation matrix from pairwise analyses and place it in the corresponding entry.
8. To assemble the K-by-K covariance matrix Σ, we recommend a three-step procedure. First, obtain correlation matrices from the estimated covariance matrices in pairwise analyses and assemble a K-by-K correlation matrix as in **Note** 7. Second, take the average of variance estimates for each tissue from pairwise analyses. Third, multiply the variance estimates with the assembled correlation matrix to get the final estimate of Σ.
9. The number of categories in the pmf increases exponentially with the number of tissues. It is generally desired to shrink small values in the pmf to zero. This will alleviate the computational burden of subsequent inferences and analyses. In practice, one could predetermine a small threshold, say 10^{-5}, and set any probability below the threshold to zero in the pmf. In empirical studies, this typically eliminates over 90% of the categories and significantly reduces the computational cost.

References

1. Nica AC, Dermitzakis ET (2013) Expression quantitative trait loci: present and future. Philos Trans R Soc B 368(1620):20120362
2. Shabalin AA (2012) Matrix eQTL: ultra fast eQTL analysis via large matrix operations. Bioinformatics 28(10):1353–1358
3. Kendziorski C, Wang P (2006) A review of statistical methods for expression quantitative trait loci mapping. Mamm Genome 17 (6):509–517
4. Gatti DM, Shabalin AA, Lam TC, Wright FA, Rusyn I, Nobel AB (2008) FastMap: fast eQTL mapping in homozygous populations. Bioinformatics 25(4):482–489
5. Cookson W, Liang L, Abecasis G, Moffatt M, Lathrop M (2009) Mapping complex disease traits with global gene expression. Nat Rev Genet 10(3):184–194
6. GTEx Consortium (2015) The genotype-tissue expression (GTEx) pilot analysis: multi-tissue gene regulation in humans. Science 348 (6235):648–660
7. GTEx Consortium (2017) Genetic effects on gene expression across human tissues. Nature 550(7675):204–213

8. Li G, Shabalin AA, Rusyn I, Wright FA, Nobel AB (2017) An empirical Bayes approach for multiple tissue eQTL analysis. Biostatistics 19 (3):391–406
9. Li G, Jima D, Wright FA, Nobel AB (2018) HT-eQTL: integrative expression quantitative trait loci analysis in a large number of human tissues. BMC Bioinformatics 19(1):95
10. Sul JH, Han B, Ye C, Choi T, Eskin E (2013) Effectively identifying eQTLs from multiple tissues by combining mixed model and meta-analytic approaches. PLoS Genet 9(6): e1003491
11. Flutre T, Wen X, Pritchard J, Stephens M (2013) A statistical framework for joint eQTL analysis in multiple tissues. PLoS Genet 9(5): e1003486
12. Lewin A, Saadi H, Peters JE, Moreno-Moral A, Lee JC, Smith KG, Petretto E, Bottolo L, Richardson S (2015) MT-HESS: an efficient Bayesian approach for simultaneous association detection in OMICS datasets, with application to eQTL mapping in multiple tissues. Bioinformatics 32(4):523–532

Chapter 17

Tissue-Specific eQTL in Zebrafish

Kimberly P. Dobrinski

Abstract

Copy number variants (CNVs) refer to the loss or gain of copies of a genomic DNA region. While some CNVs may play a role in species evolution by enriching the diversity of an organism, CNVs may also be linked to certain diseases such as neurological disorders, early onset obesity, and cancer. CNVs may affect gene expression by direct overlap of the genic region or by an indirect effect where the CNV is located outside the gene location. These indirect CNV regions may contain regulatory elements such as transcription enhancers or repressors as well as regulators such as miRNAs which may work at the level of transcription or translation. *Danio rerio* (zebrafish) is an excellent model organism for CNV studies. Zebrafish genomes contain a large amount of variation with 14.6% of the zebrafish reference genome found to be copy number variable. This level of variation is more than four times the percentage of reference genome sequence covered by similarly common CNVs in humans. It is this high level of variation that makes zebrafish interesting to investigate the effects of CNV on gene expression. Additionally, zebrafish share 70% of genetic similarities with humans, and 84% of genes associated with human disease are also found in zebrafish. Expressive quantitative trait loci (eQTL) analysis may be used in zebrafish to explore how CNVs may affect gene expression in both a direct and indirect manner. eQTL analysis may be performed for cis associations with a 1-Mb (megabase) window upstream and downstream from the transcription probe midpoint to CGH midpoint. Trans associations (variants that are located beyond the 1-Mb window of the gene either on the same chromosome as the gene or on a different chromosome) may be investigated as well through eQTL analysis; however, trans associations require more tests to be performed than cis associations, which limits power to detect associations. Pairwise associations between each pair of copy number variant and gene will be investigated separately from the same individual using Spearman rank correlations with significant associations found being followed with a multi-test correction technique to assess significance of those CNV gene expression associations. An association between a CNV to a gene expression phenotype should be considered significant only if the p value from the analysis of the observed data is lower than the 0.001 tail threshold from a distribution of the minimal p values (which are found from all comparisons for a given gene from 10,000 permutations of the expression phenotypes). Associations between CNVs and genes may be found to be direct or indirect as well as positive (increased copy number—increased expression) or negative (increased copy number—decreased expression, decreased copy number—increased expression). Ongoing analyses with these associations will investigate the impact of CNVs on gene functionality including immune function and potential disease susceptibility.

Key words Copy number variant, Structural variation, Expression quantitative trait loci (eQTL), Transcription regulation, Zebrafish, Pathogenesis

Xinghua Mindy Shi (ed.), *eQTL Analysis: Methods and Protocols*, Methods in Molecular Biology, vol. 2082, https://doi.org/10.1007/978-1-0716-0026-9_17,

1 Copy Number Variation

Genome variation such as single nucleotide polymorphisms (*SNP*) and variable numbers of tandem repeats (*VNTR*) has been studied extensively while copy number variations (*CNV*), a more recent discovery, are now regarded as an important source of genetic variation in humans and animals [1, 2]. Structural variants include balanced (most inversions, insertions, translocations) and unbalanced rearrangements such as CNVs which represent the largest known component [3–9]. CNVs are found across genomes and may range from kilobases (kb) to several megabases (Mb) [3]. While researchers believe some CNVs enrich the diversity of an organism and play a role in species evolution [1–3], CNVs may also be linked to certain diseases such as neurological disorders [10], early onset obesity [11], and cancer. An increase in copies of certain genes or genomic regulatory regions could lead to rapid cell growth or heightened gene expression of oncogenes causing disease [12, 13]. Therefore, it is important to have a thorough understanding of CNVs across the genome and the role they may play in regulation of gene expression.

2 Zebrafish as a Model System

The zebrafish, *Danio rerio*, is utilized as a model system for human development as well as disease such as cancer. Zebrafish are less expensive than mouse models to maintain, while sharing 70% of genetic similarities with humans [14]. Additionally, 84% of genes associated with human disease are also found in zebrafish, making this model well suited for human disease studies [14]. Additional benefits include its easiness to breed, external fertilization, and the availability of forward and reverse genetic techniques [15]. Zebrafish are an excellent human model; however, genomic variants such as single-nucleotide polymorphisms and structural genomic variants have not been well classified in the zebrafish genome. Contributions to this delay may have included initial problems with the first zebrafish genome assemblies, including a high repeat frequency and a high degree of polymorphism relative to mammals [16]. Additionally, starting material used for the BAC and shotgun libraries were derived from several non-isogenic fish. An additional fosmid library was then prepared from a single double-haploid Tubingen (Tu) fish [16]. It is now known, however, zebrafish genomes contain a large amount of variation with the generation of a CNV zebrafish map discovering a nonredundant dataset comprising of 192,460,331 bp of sequence, representing 14.6% of the zebrafish reference genome [17]. This level of variation is more than four times the percentage of reference genome sequence covered by similarly common CNVs in humans [3] and other vertebrates [4, 7, 8]. CNV elements from this study (CNVEs; e.g., CNVs having >50% reciprocal overlap) exhibited strain specificity with 69% of CNV elements unique to one strain, with the highest levels observed for Tu [17]. Additionally, this study indicated that only 37% of CNVEs overlapped National Center for Biotechnology Information (NCBI) Reference Sequence (RefSeq) genes which represented a significant depletion in the number of CNVEs expected to overlap RefSeq genes by chance alone [17]. The newest zebrafish genome build generated by the Genome Research Consortium, GRCz11, was released May of 2017.

3 eQTL with Copy Number Variants in Zebrafish

One way to complete eQTL analysis in zebrafish is to complete a pairwise association utilizing correlation or regression analysis between the genotype (using probe intensity values across the variant) and the expression of individual genes across different individuals. The assumption in this type of analysis is that genetic variants are independent of each other and gene expression should also be independent among the genes. With these assumptions, completion of association analysis on each of the variant and gene pairs may be carried out with additional multi-test corrections performed to account for the bias introduced by the large number of individual tests.

One study in humans discovered 3534 genes affected by eQTL in cis (using variants within a specific window of gene location) and 48 genes affected in trans [18]. Determination of cis and trans is mainly based on the distance boundary between genetic variants and the target gene. Cis associations may be focused on genetic variants that are found within a certain distance from either the midpoint or the transcription start site of each gene. However, it must be remembered that trans-eQTLs may have weaker effect sizes than cis-eQTLs [18]. Additionally, this same study found that CNVs are more likely to be eQTLs than SNPs [18]. An initial zebrafish cis-eQTL study of seven full sib adult zebrafish tested $\log_2$ ratios of 15,137 CNV probes against expression levels of 11,953 gene probes and found 301 significant CNV probe–gene probe expression associations comprised of 232 CNVs (29.3% of those tested) and 255 genes (2.5% of known genes) [17]. The following associations were found: 25.2% direct positive associations, 13.3% direct negative associations, 34.6% indirect positive associations, and 26.9% indirect negative associations with direct indicating gene overlap and indirect indicating the variant is outside of the gene. Interestingly, the majority of CNV–gene expression associations were indirect (61.5%) and therefore possibly regulatory in nature [17].

4 Copy Number Variation Discovery via Array Comparative Genomic Hybridization

To carry out eQTL analyses for tissue-specific data, tissues should be collected from adult zebrafish. Tu zebrafish have a high level of CNVs and therefore may provide a good model to use for eQTL studies. Array comparative genomic hybridization (aCGH) can be used to generate CNV calls. aCGH can be performed using a custom-designed Agilent Technologies SurePrint G3 CGH microarray which maps 60-mer oligonucleotides generated with an algorithm from the zebrafish reference genome. Unique probes should be selected to reduce noise generated from highly repetitive regions (i.e., segmental duplications, LINEs). In the previously used array, a total of 389,813 features were tiled across the zebrafish genome along with positive and negative controls on the array. This design provides an average 2.9-kb probe spacing throughout the zebrafish reference genome. For aCGH experiments, one Tu individual is chosen randomly and used as a reference sample to compare against all other individuals (test samples, Fig. 1a). Zebrafish aCGH carried out on separate tissues do not indicate CNV mosaicism across tissues, but instead demonstrate consistent DNA make-up across tissues within healthy individuals. Additionally, self-self hybridizations carried out previously in these fish only indicated 30 self-self calls which were removed from future analyses. Therefore, whole fish or a single tissue can be used to generate CNV calls. To carry out aCGH, DNA extracted from samples are labeled with the Bioprime system per manufacturer instructions (Invitrogen) and hybridized using the standard Agilent aCGH protocol with the following modification: 500 ng of heat-denatured DNA (5 min at 95 °C) is best used per labeling reaction in zebrafish in place of the 1 μg of restriction enzyme-digested DNA. Once hybridization is complete, arrays should be scanned on an Agilent G2565CA Microarray Scanner with SureScan High Resolution Technology containing lasers specific for cy3 and cy5 fluorescence and then use a 2 μm resolution. Next, images may be extracted using Agilent Feature Extraction software (Agilent Technologies, Inc., Santa Clara, CA) incorporating signal normalization for the Cy3 and Cy5 signal intensities across the array. To determine copy number variable regions, Nexus Copy Number software (Biodiscovery, Inc., El Segundo, CA) may be used to generate CNV calls based on $\log_2$ ratio output files using a rank segmentation algorithm. Settings should be optimized using self-self hybridizations to reduce false positive calls. Array settings should be set as follows: Significance Threshold 1.0×10^{-5}, Maximum Continuous Probe Spacing 200 kb, Minimum Number of Probes per Sequence 3, High Gain 1.0, Gain 0.4, Loss −0.4, Big Loss −0.8, and with no outliers removed. As mentioned previously, these settings produced 30 self-self calls which were removed from the analyses. Once

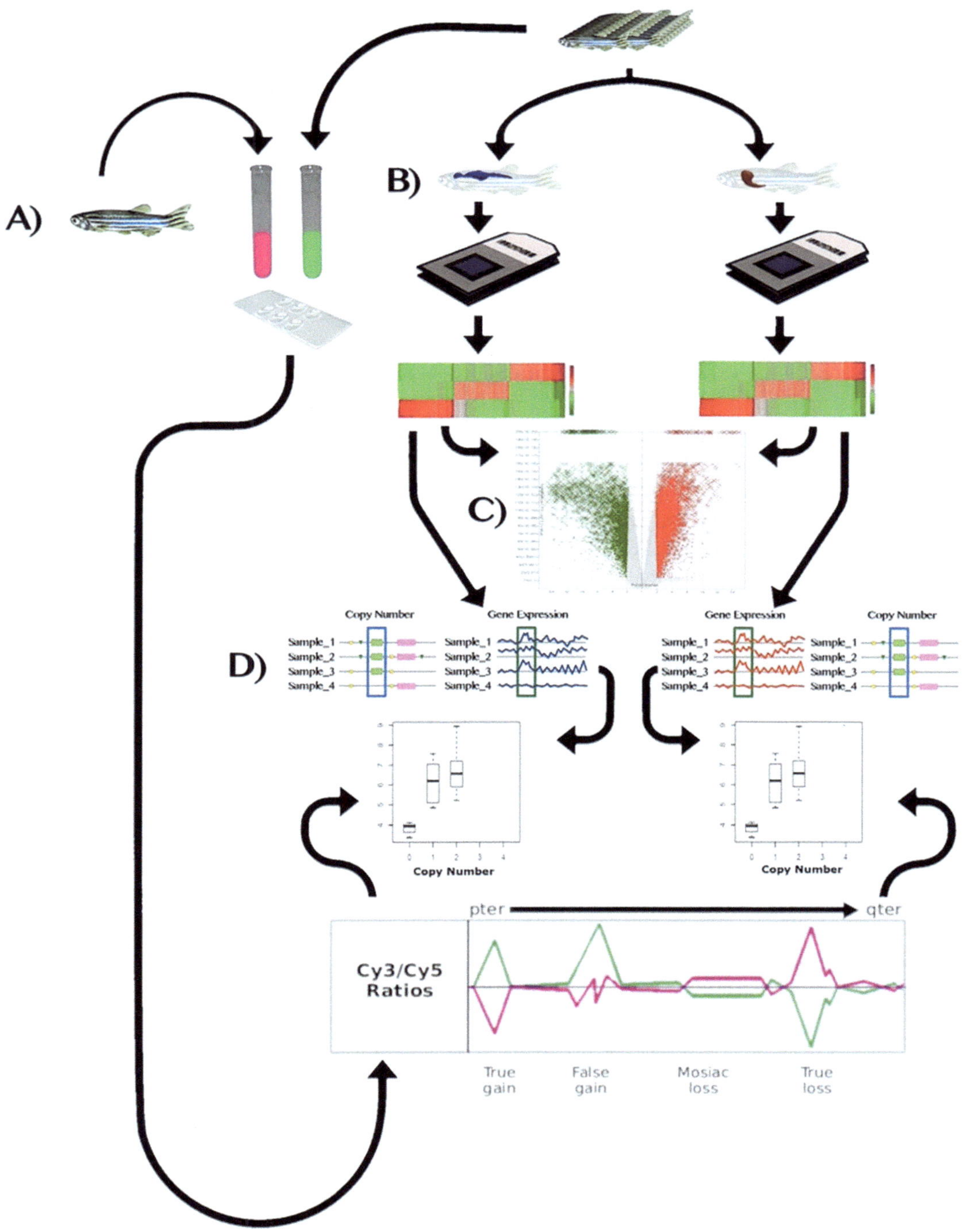

Fig. 1 Experimental flowchart of tissue-specific eQTL in zebrafish. (**a**) Array comparative genomic hybridization between 1 reference fish and 30 sample fish. (**b**) Tissue-specific gene expression using zebrafish expression arrays. (**c**) Differential expression analysis for all zebrafish genes between liver and kidney tissues. (**d**) Tissue-specific pairwise cis-eQTL analysis for genes and copy number variants located within a 1 MB window of gene midpoint

CNV calls are generated, a 50% reciprocal overlap is carried out across all individuals in the study to generate CNVEs. This overlap is carried out using custom PERL scripts combining calls sorted by size (smallest to largest), and then merging those which overlap each other by at least 50% of their respective total lengths.

5 qPCR Validation

Prior to eQTL, CNV regions found by aCGH should be confirmed via qPCR validation. Sequences for CNV regions should be obtained from the zebrafish genome sequence (http://genome.ucsc.edu/). Primers for an ultra-conserved element (UCE) may be used as control primers, (5′–3′) (F) CCTTTTCCGATGCTTTTTACAC and (R) GGAAGCCTAGTGCAGTGCTAGT. The 10-μL qPCR assays are performed in triplicate using Fast SYBR® Green Master Mix (Applied Biosystems) in a Step One Plus Real Time PCR System (Applied Biosystems) with incubation parameters 95 °C for 10 min, followed by 35 cycles of 95 °C for 15 s and 60 °C for 30 s. Dissociation curves will be generated at the conclusion of the reaction and can be used to confirm the specificity of PCR products. Fold differences in DNA amount between tests and reference sample are calculated as $2^{-\Delta\Delta CT}$, where $\Delta\Delta CT = [(CT\ target - CT\ UCE) - (CT\ target_{Ref} - CT\ UCE_{Ref})]$; CT target − CT UCE are the CT values for target and UCE amplification in tests, and $CT\ target_{Ref} - CT\ UCE_{Ref}$ are the corresponding values from the reference sample. Primers should be added to qPCR at 10 μM with 3 μL of cDNA. The $2^{-\Delta\Delta CT}$ for each CNV can then be compared with the $\log_2$ ratio obtained from aCGH.

6 Expression Analyses

For expression analyses, sense-strand cDNA is generated from total RNA (50 ng) from liver and kidney tissues for each fish with the Ambion® WT Expression Kit (4411973), followed by fragmentation and labeling with the GeneChip® WT Terminal Labeling and Controls Kit, (901524, Affymetrix), and placed on Zebrafish Gene 1.0 ST expression arrays (902007, Affymetrix) for hybridization (Fig. 1b). This array has whole-transcriptome coverage based on the ZV8 zebrafish genome build and utilizes 25-mer probes providing for up to 22 probes across the full length of each gene for a total coverage of 550 bases per transcript. The array contains 1,255,682 probes for a total of 59,302 gene level probe sets. Following hybridization, slides are washed on a GeneChip Fluidics Station (Affymetrix) as directed and read using an Affymetrix Microarray Scanner. Affymetrix Expression Console software is used for analysis and normalization utilizing the RMA (Robust Multiarray Averaging) algorithm to obtain an intensity value signal for each probe set. Affymetrix® Transcriptome Analysis Console (TAC) is used to determine tissue-specific expression differences. TAC generates a *p* value from ANOVA, which considers both fold change and variability. To correct for the large number of multiple comparisons involved in microarray analysis, TAC also provides a more stringent false discovery rate calculation based on Benjamini-Hochberg step-up FDR-controlling procedure and provides an adjusted *p* value. A 5% FDR is used for significant *p* values for differential expression analysis (Fig. 1c).

7 qRT-PCR for Validation of Expression Profiles

Prior to carrying out eQTL analyses, a subset of genes showing differential expression across tissues can be confirmed by utilizing the following protocol. Total RNA (3.7 μg) should be treated with RQ1 RNase-Free DNase (Promega) for 30 min at 37 °C. DNase is then inactivated by adding 1 μL of stop solution and 10 min incubation at 65 °C. DNase-treated RNA (3 μL) will be reverse transcribed using GeneAmp® RNA PCR Core Kit (Applied Biosystems). Prior to enzyme addition, samples will undergo a 3 min incubation at 72 °C and then reverse transcription may be carried out in a GeneAmp PCR System 9600 (Perkin Elmer) 25 °C for 10 min, 42 °C for 60 min, 95 °C for 10 min, and 4 °C for cooling. Additionally, each sample should be prepared in duplicate, with one sample lacking reverse transcriptase for a negative control. After cDNA generation, qPCR amplification is carried out using Fast SYBR® Green Master Mix (Applied Biosystems) in a Step One Plus Real Time PCR System (Applied Biosystems). Incubation parameters for qPCR should include: 50 °C for 2 min, 95 °C for 10 min, then 40 cycles of 95 °C for 15 s and 60 °C for 1 min. Fold differences in transcription between liver and kidney are calculated as $2^{-\Delta\Delta CT}$, where $\Delta\Delta CT = [(CT\ \text{target} - CT\ \beta\text{-actin}) - (CT\ \text{target}_{Ref} - CT\ \beta\text{-actin}_{Ref})]$; CT target − CT β-actin are the CT values for liver, and $CT\ \text{target}_{Ref} - CT\ \beta\text{-actin}_{Ref}$ are the corresponding values from kidney. Primers should be designed based on the sequences of interest and added to qPCR at 10 μM with 3 μL of cDNA.

8 eQTL in Adult Zebrafish Tissues

To compensate for tissue-specific expression, if whole fish are being used for transcription analyses and not specific tissues, then it is best to focus only on associations between genes and CNV regions that demonstrate homozygous copy number loss (i.e., copy number of zero) and high copy gain (i.e., a copy number greater than 4). Expression quantitative trait loci (eQTL) analyses are performed for cis associations with a 1-Mb (megabase) window upstream and downstream from the transcription probe midpoint to CGH midpoint (Fig. 1d). If tissues are being analyzed, then there is no need to focus on only homozygous losses and high copy gains. Pairwise associations between each pair of copy number variant and gene within liver and kidney tissues are investigated separately from the same individual using Spearman rank correlations [19]. All genes included in the analysis should have a *p* value cutoff of 0.05. Additionally, CNVEs included in eQTL analysis should be found in at least two fish. Once Spearman rank correlations have been completed, each pair will have a nominal *p* value and correlation value. Afterward, a multi-test correction technique must be performed to select significant associations. In order to assess significance of CNV gene expression associations, 10,000 permutations of each expression phenotype relative to the CGH-$\log_2$ ratios should be performed. An association to a gene expression phenotype should be considered significant if the *p* value from the analysis of the observed data is lower than the 0.001 tail threshold from a distribution of the minimal *p* values. These *p* values will come from all comparisons for a given gene from 10,000 permutations of the expression phenotypes [19–21].

Although many eQTL studies have shown SNP expression associations [22–30], more recent studies have begun to focus on CNV eQTL analysis [31–33]. These studies indicate that CNVs contribute significantly to gene expression variation. The identified associations can be visualized in box plots (Fig. 1d), where the *X* axis represents the copy number for an associated variant with the *Y* axis representing gene expression values associated with the variant for each tissue within those same individuals. Copy number effects can be negative or positive and would subsequently be represented by a negative or positive correlation value. Trans associations may also be carried out to assess the copy number variants that are located beyond the 1-Mb window of the gene either on the same chromosome as the gene or on a different chromosome. However, it must be remembered that trans associations require more tests to be performed than cis associations, and therefore the burden of multiple testing is increased for trans associations thus limiting the power to detect associations.

9 Principle Component Analysis

Batch effects are often seen with aCGH and expression data. These effects may be due to technical variables, different lots of reagents, genetic variables such as GC content, or perhaps environmental variables. Normalization is carried out after feature extraction for aCGH arrays by using combined rank consistency filtering with LOWESS intensity normalization. Each array contains spike in controls. Additionally, Agilent's feature extraction software provides a QC report for every array scan reporting signal intensity and signal-to-noise values. This allows problems with specific arrays to be tracked and these data can be omitted from the study. For expression arrays, spiked in controls should maintain a 1:2 ratio between the 5′ and 3′ probe sets. There is a measure of background noise which must remain constant across all arrays. The values obtained for background noise (RawQ) should be within ±3 points of the median. Box plots and histograms of each array are performed. Histograms will identify saturation. Probes with the highest intensities in the plot should be removed from the analyses. Box plots have the median intensity within the box with 25th and 75th percentiles as the outline of the box. Whisker lines indicate the spread of the data. Any arrays that are not similar in range are discarded.

In addition to these normalization and QC methods, principal components analysis (PCA) may be used to identify heterogeneity across arrays within the study. Any principal components that are identified in the study as confounders must be removed or measured as covariates within the eQTL analyses.

10 Conclusions

Zebrafish remain an excellent model for human development and disease. Recent CNV studies indicate there is a large amount of variation in the zebrafish, more than four times what has been seen with human studies [17]. Additionally, initial eQTL studies in zebrafish indicate the majority of gene-variant associations are indirect associations [17]. These two factors make zebrafish a very interesting model to investigate copy number variation and the effects of CNV on gene regulation. With the latest genome build recently finished and new tools being developed to investigate the zebrafish genome, this organism is at the forefront of being utilized to design eQTL studies with results that can be potentially useful for translation to human disease.

Acknowledgments

Jason Dobrinski for his assistance with Fig. 1.

References

1. Beckmann JS, Estivill X, Antonarakis SE (2007) Copy number variants and genetic traits: closer to the resolution of phenotypic to genotypic variability. Nat Rev Genet 8:639–646
2. Yim SH, Kim TM, Hu HJ, Kim JH, Kim BJ, Lee JY, Han BG, Shin SH, Jung SH, Chung YJ (2010) Copy number variations in East-Asian population and their evolutionary and functional implications. Hum Mol Genet 19:1001–1008
3. Conrad DF, Pinto D, Redon R et al (2010) Origins and functional impact of copy number variation in the human genome. Nature 464:704–712
4. Egan CM, Sridhar S, Wigler M, Hall IM (2007) Recurrent DNA copy number variation in the laboratory mouse. Nat Genet 39:1384–1389
5. Iafrate AJ, Feuk L, Rivera MN, Listewnik ML, Donahoe PK, Qi Y et al (2004) Detection of large-scale variation in the human genome. Nat Genet 36:949
6. Kidd JM et al (2008) Mapping and sequencing of structural variation from eight human genomes. Nature 453(7191):56–64
7. Lee AS et al (2008) Analysis of copy number variation in the rhesus macaque genome identifies candidate loci for evolutionary and human disease studies. Hum Mol Genet 17:1127–1136
8. Perry GH et al (2006) Hotspots for copy number variation in chimpanzees and humans. Proc Natl Acad Sci U S A 103:8006–8011
9. Redon R et al (2006) Global variation in copy number in the human genome. Nature 444:444–454
10. Glessner JT et al (2009) Autism genome-wide copy number variation reveals ubiquitin and neuronal genes. Nature 459:569–573
11. Bochukova EG et al (2010) Large, rare chromosomal deletions associated with severe early-onset obesity. Nature 463:666–670
12. Shlien A, Malkin D (2009) Copy number variations and cancer. Genome Med 1(6):62
13. Chen EY, Dobrinski KP, Brown KH, Clagg R, Edelman E, Ignatius MS et al (2013) Cross-species array comparative genomic hybridization identifies novel oncogenic events in zebrafish and human embryonal rhabdomyosarcoma. PLoS Genet 9(8):e1003727
14. Howe K, Clark M, Stemple D (2013) The zebrafish reference genome sequence and its relationship to the human genome. Nature 496:498–503
15. Lieschke GJ, Currie PD (2007) Animal models of human disease: zebrafish swim into view. Nat Rev Genet 8:353–367
16. Carpio Y, Estrada M (2006) Zebrafish as a genetic model organism. Biotecnol Apl 23:4
17. Brown KH, Dobrinski KP, Lee AS, Gokcumen O, Mills RE, Shi X et al (2012) Extensive genetic diversity and substructuring among zebrafish strains revealed through copy number variant analysis. Proc Natl Acad Sci U S A 109(2):529–534
18. Bryois J, Buil A, Evans DM, Kemp JP, Montgomery SB, Conrad DF, Ho KM, Ring S, Hurles M, Deloukas P, Davey Smith G et al (2014) Cis and trans effects of human genomic variants on gene expression. PLoS Genet 10(7):e1004461
19. Stranger B, Forrest M et al (2007) Relative impact of nucleotide and copy number variation on gene expression phenotypes. Science 315(5813):848–853
20. Schadt E et al (2003) Genetics of gene expression surveyed in maize, mouse and man. Nature 422:297
21. Chesler EJ, Lu L et al (2005) Complex trait analysis of gene expression uncovers polygenic and pleiotropic networks that modulate nervous system function. Nat Genet 37(3):233–242
22. Stranger B, Forrest M et al (2005) Genome-wide associations of gene expression variation in humans. PLoS Genet 1(6):e78
23. Benjamini Y, Hochberg Y (1995) Controlling the false discovery rate: a practical and powerful approach to multiple testing. J R Stat Soc Ser B 57(1):289–300
24. GTEx Consortium (2013) The genotype-tissue expression (GTEx) project. Nat Genet 45(6):580–585
25. Stranger B, Montgomery S et al (2012) Patterns of cis regulatory variation in diverse human populations. PLoS Genet 8(4):e1002639
26. Li Q, Seo J, Stranger B et al (2013) Integrative eQTL-based analyses reveal the biology of breast cancer risk loci. Cell 152(3):633–641
27. Pickrell J, Marioni JC et al (2010) Understanding mechanisms underlying human gene expression variation with RNA sequencing. Science 464(7289):768–772
28. Liang L, Morar N et al (2013) A cross-platform analysis of 14 177 expression quantitative trait

loci derived from lymphoblastoid cell lines. Genome Res 23(4):716–726

29. Kreimer A, Pe'er I (2013) Variants in exons and in transcription factors affect gene expression in trans. Genome Biol 14(7):R71
30. Cheung V, Nayak R, Wang I, Elwyn S, Cousins S, Morley M, Spielman R (2010) Polymorphic cis and trans-regulation of human gene expression. PLoS Biol 8(9):e1000480
31. Schlattl A, Anders S, Waszak SM, Huber W, Korbel JO (2011) Relating CNVs to transcriptome data at fine resolution: assessment of the effect of variant size, type, and overlap with functional regions. Genome Res 21 (12):2004–2013
32. Montgomery S, Sammeth M, Gutierrez-Arcelus M et al (2010) Transcriptome genetics using second generation sequencing in a Caucasian population. Science 464 (7289):773–777
33. Tian L, Quitadamo A, Lin F, Shi XM (2014) Methods for population-based eQTL analysis in human genetics. Tsinghua Sci Technol 19:624–634

INDEX

Xinghua Mindy Shi (ed.), *eQTL Analysis: Methods and Protocols*, Methods in Molecular Biology, vol. 2082,
https://doi.org/10.1007/978-1-0716-0026-9,

MIX
Papier aus verantwortungsvollen Quellen
Paper from responsible sources
FSC® C105338

If you have any concerns about our products,
you can contact us on
ProductSafety@springernature.com

In case Publisher is established outside the EU,
the EU authorized representative is:
Springer Nature Customer Service Center GmbH
Europaplatz 3, 69115 Heidelberg, Germany

Printed by Libri Plureos GmbH
in Hamburg, Germany